Soil Fertility and Nutrient Management in Horticulture

NIPA® GENX ELECTRONIC RESOURCES & SOLUTIONS P. LTD.
New Delhi-110 034

www.nipaers.com
Browse, Search, Read & Buy...
Print Books
eChapters
eBooks
Publishers
Forthcoming
Subject Catalogues
New Titles
Online Resources on:
✓ Current Affairs
✓ Reasoning, Logic & Aptitude
✓ Competitive Examinations
✓ English Language Lab
✓ Writing & Pronunciation Tools
✓ Personality Development
✓ Online Programmes for Professionals
✓ Interview Preparation
Pay Using
PhonePe
Google Pay
paytm
PayPal
UPI
CCAvenue

Soil Fertility and Nutrient Management in Horticulture

Chittaranjan Sarangi, Ph.D.
Assistant Professor (Soil Science)
College of Horticulture (OUAT), Chiplima
Odisha, India

Gour Hari Santra
Professor (Soil Science)
College of Agriculture
Orissa University of Agriculture & Technology
Bhubaneshwar, Orissa, India

Biswanath Sahoo
Scientist (Horticulture)
Krishi Vigyan Kendra (OUAT) Bhadrak
Orissa, India

NIPA® GENX ELECTRONIC RESOURCES & SOLUTIONS P. LTD.
New Delhi-110 034

NIPA® GENX ELECTRONIC RESOURCES & SOLUTIONS P. LTD.

101,103, Vikas Surya Plaza, CU Block
L.S.C. Market, Pitam Pura, New Delhi-110 034
Ph : +91 11 27341616, 27341717, 27341718
E-mail: newindiapublishingagency@gmail.com
web: www.nipabooks.com

For customer assistance, please contact
Phone: + 91-11-27 34 17 17
Fax: + 91-11- 27 34 16 16
E-Mail: feedbacks@nipabooks.com

ISBN: 978-81-19254-35-4

Composed and Designed by NIPA®.

Dedicated to
The Seekers of Knowledge

Preface

Agricultural, horticultural, forestry and fodder production *etc.* for the burgeoning population are ever increasing enterprises that impose heavy tasks on the management of natural resources. Short term profitability at the cost of sustainability disharmonises the natural forces and processes resulting in peril of the inhabitants of the earth. To this end an understanding with conviction of the interdependency of processes will enhance the realisation of goals in harmony with nature.

Intended primarily to be a text book, it will help to educate learners to follow those principles. Students, planners, faculty and any one interested in achieving those goals will find it necessarily useful.

We take responsibility for any mistake. The readers are requested to communicate any such mistakes which will be duly taken care of.

Authors

Acknowledgement

We deeply acknowledge the work itself that it could come to an end. The endeavour put by various authors from which we have drawn freely has motivated us to undertake such a task is fully acknowledged.

We are thankful to the students teaching whom was an enlightened experience that stimulated us to reach wider audience through this work.

We are very much thankful to M/S New India Publishing Agency, New Delhi who have made it possible to bring the work in book form for the benefit of the readers.

The co-operation by each and everyone extended is cordially acknowledged.

Authors

Contents

PART - I : SOIL FERTILITY

Plant Nutrients-Importance in plant nutrition, Processes and factors governing their Availability from Soil.

Chapter 1

Introduction to Soil Fertility and Productivity

Soil fertility -importance

From edaphological point of view soil is a medium for plant growth. It holds and nourishes the plants. Soil fertility refers to the ability of soil to supply plant nutrients. Soil fertility and fertilizers are very much closely related terms; and cannot be neglected one from the other in their function. Soil fertility acts as a 'SINK' where in plants can draw nutrients for maximum productivity, where as fertilizer, acts as a 'SOURCE' wherein we can apply continuously different nutrients. The importance of soil fertility and nutrient management is being increasingly recognized in all countries recently to meet the demand for food and other agricultural raw materials. Intensive use of fertilizer, intensive cropping and high yielding seeds have no doubt increased the food production and reduced the food shortage but it has also ushered in innumerable problems of soil fertility, soil and water pollution. On the other hand, fast depletion of nutrients due to over exploitation, a wide spread deficiency of N, P, K and S coupled with micronutrients deficiencies specially Zn, B and Mo is taking place. Further deforestation, shifting cultivation, burning of crop residues, trees, bushes, grasses and cow dung, soil erosion, nutrient losses and imbalanced nutrient application, leaching losses *etc*., have aggravated the depletion of soil fertility status. The productivity of Indian agriculture solely depends on scientific management of soil fertility along with judicious and efficient use of fertilizers integrated with organic sources.

Soil fertility problems can be solved not only by supply of plant food nutrients, but their efficient management since the fertilizer is one of the costliest inputs. A well balanced scheduling of fertilizers is a scientific tool to get maximum returns with minimum investment. Apart from fertilizers, lack of biomass resources, and consequent supply of organic manures add to the severity. So with all these conditions soils become depleted and very "hungry" for the nutrients with time. It is therefore, imperative that sound soil and crop management practices, Judicious use of fertilizers and Integrated nutrient management practices must be devised scientifically to improve and maintain good soil fertility and better soil productivity for profitably sustained crop production.

Historical developments of soil fertility

The concept of soil fertility and its management to improve crop yields is is as old as the development of agriculture. In ancient times, they had knowledge of applying manures such as farm yard manure, green manure, night soil, bone, wood ashes, *etc*., to soil for the purpose of increasing crop yields.

Xenophan (430-355 B.C), a Greek historian first recorded the merits of green manure crops, Cato (234-149 B.C) wrote a practical hand book and recommended intensive cultivation, crop rotations, and the use of legumes for livestock farming. He was first to classify "Land" based on specific crops. Columella (A.D. 45) emphasized the usefulness of turnips for soil improvements. He also advocated land drainage and the use of ashes, marl (lime deposits), clover and alfalfa to make the soil more productive. JethroTull (1731) and Francis Home (1757) claimed that Nitre (Nitrate Salts), water, air, earth, Epsom salt ($MgSO_4$), Saltpetre (Sodium & Potassium Nitrate), Vitriolated tarter (Potassium sulfate) and Olive oil increased plant growth.

The modern concepts of mineral nutrition has been developed relatively recently during the last one and half century. In the early 19th century two prominent scientists, Nicholas Theodore de Seussure (1804), a Swiss Physicist and Jean Baptiste Boussingault (1834), a French Chemist & Agriculturist, were first to report that plants need mineral nutrients for growth and development. J.B. Boussingault started field plot experiments on his farm. Justus Von Liebig (1840), a German Chemist, reported growing plants obtain elements Ca, K, S and P from the soil, whereas carbon is obtained from CO_2 in the air and not from the soil. He also suggested that plants obtain H & O from air as well as from water and N from ammonia. He also established certain basic principles of sound soil management-A cropped soil is restored to fertility only by adding to it all minerals & N removed by the plants, established the theory of "Law of Minimum" in relation to plant nutrition. The law states that the productivity of a crop is decided by most limiting factor. He is regarded as the "Father of Agricultural Chemistry" John B. Lawes (1837) of the Rothamsted Experiment station, England was first to make and use Super phosphate on his farm (1840). Both J.B. Lawes & J.H. Gilbert (1852) applied the principles of Liebig and stated that addition of mineral fertilizers to cropped soils would keep the soil fertile. They further elaborated the chemistry of plant nutrition. A. Gris (1844) discovered that the

'Chlorosis' of some plants can be corrected by sprays of iron salt and demonstrated its essential nature in plant nutrition. By 1860, German Botanist Julius Von Sachs and others, established the essentiality of 10 nutrient elements and were using these elements in the synthetic mineral nutrient solutions for the growth of plant. Discovery of other five nutrients was made after more than 60-70 years, while the last one in the group "Chlorine" has established to be essential for plant growth only in 1954.

Soil fertility

It is defined as the inherent capacity of a soil to supply essential nutrients to plants in an adequate amount and in suitable proportions to maintain their growth and development.

Factors affecting soil fertility

I. Natural factors or Pedogenic factors

II. Edaphic factors or Soil management factors

I. **Natural Factors:** Influence the formation of the Soil. These are

a. Parent material

b. Climate and vegetation

c. Topography and age of soil

a. ***Parent material***:It is the important raw materials for the formation of any soils. If the parent material is rich in plant nutrients, the soils formed from it are usually fertile. The property of soil depends on the property of parent rock . As for example,

i. Sandy loam soils, formed from Granite and granite gneiss are low in fertility,

ii. Lime stone and Basalt rock which are easily weathered results fine textured, very fertile and dark coloured soil.

iii. Soils derived from calcareous rock contain more P than the soil derived from granite.

iv. Sand stone-leads to coarse textured, sandy soil of low fertility.

v. Shale- forms clayey soil, but not very fertile.

b. ***Climate and vegetation***: These are interrelated factors as the amount and type of vegetation in an area depends on climate especially rainfall and temperature. These two factors in turn influence the type of soil fertility.

Under hot and humid region, the natural vegetation is forest, which develops more fertile soils due to accumulation of forest litter and organic matter. In semi arid conditions the natural vegetation is grass which leads to more accumulation of organic matter in soil surface layer due to fibrous root system, and good soil aggregation. These soils are more fertile than the area under forest vegetation.

Temperate soils are very fertile compared to tropical soils because of the lesser decomposition of organic matter in soil due to very low temperature. On the other hand the tropical soils are less fertile due to intense weathering, hyperthermic temperature regime coupled with high microbial activity resulting in faster rate of disintegration and decomposition of organic matter in the tropics.

c. ***Topography and age of soil*:** The soils of hilly tracts are usually poor because of excessive leaching and erosion of the top soil. In sloppy land, the soils of low lying areas are usually richer because of the transportation and accumulation of soil and plant nutrients. Similarly, old soils are less fertile due to excessive weathering, leaching and continuous cultivation.

II. **Edaphic factors or soil management factors:** Include the soil management practices that are concerned with addition or removal of plant nutrients.

1. ***Physical conditions of soil***

a. **Texture of soil**: Fine textured soils have greater surface area, greater CEC and so better soil fertility than the coarse textured soils.

b. **Structure of soil**: Well aggregated soils are more productive compared to non aggregated soils as the pores act as storage of water and air.

c. **Soil water**: Clayey soils hold more water than sandy soils, hence they are more productive

d. **Soil aeration**: Soil air containing oxygen is essential for root and microbial respiration, decomposition of soil organic matter and uptake of nutrients by plants. Higher CO_2 content in the soil restricts the uptake of nutrients. Soil aeration determines oxidation and reduction process of soils.

e. **Soil temperature**: It is required for metabolic activity of plants, microbial activity and decomposition process. Temperature variations affect the nutrient absorption and nutrient conversions in soil and ultimately plant growth.

f. **Soil compaction and tillage operations**: Compactness decides the aeration status of soil and have direct effect on the ability of plant roots to absorb both nutrients and moisture from the soil. Tillage operations using heavy implements will destroy the good soil structure which intern affect the soil fertility status.

2. ***Chemical conditions of soil***
   a. **Soil reaction (pH)**: Availability of nutrients in soils is greatly influenced by change in soil pH. The neutral pH of 6.5-7.5 is optimum for good productive soils.
   b. **Salt content (EC)**: Salinization affects water relations of plants because of osmotic pressure.
   c. **Microorganisms**: Soil microorganisms improve the soil fertility as they decompose organic matter and mineralize nutrients in soil. They are also involved in nutrients cycling by mineralization, fixation, absorption and solubilization of nutrients in soils.
3. ***Root growth and extension***: Root performs absorption of water and nutrients needed for the plant. Root metabolism creates a nutrient demand. Dense and extensive root system helps better nutrient foraging.
4. ***Organic matter content of the soil*:** The more the organic matter content, the higher is the fertility status. Organic matter increases humus content and CEC of soils. It improves the physical properties of soil like structure, good aggregation of soil particles, aeration, and water holding capacity, solubility of the minerals and supplies "food and energy" for the growth and development of microorganisms.
5. ***Cropping system:***Cultivation of same crop continuously in the same field without replenishment decreases the soil fertility. Thus, inclusion of various crops and cropping systems like double, mixed, relay, multiple cropping and crop rotation increases the soil fertility.
6. ***Soil erosion:*** Erosion is the physical removal of top soil by water and wind. As such it decreases the soil fertility and promotes soil degradation due to nutrientloss.

Soil Productivity

According to modem usage, soil fertility is the capacity of the soil to produce crops of economic value and to maintain health of the soil without deterioration.Soil productivity is the crop producing capacity of a soil which is measured in terms of yield (bio-mass). It has an economic connotation. Productivity is a very broad term and fertility is only one of the factors that determine the crop yields. To be productive, soil must contain all the 14 essential nutrients required by the plants. The total quantity of nutrients is not only being sufficient but they should also be present in an easily "available" form and in "balanced" proportions. Over and above fertility, there are other factors deciding productivity. Productivity of soil is crop specific. Water logged soils are productive for rice and unproductive for aerobic crops.

Differences between soil fertility and productivity

Soil Fertility	Soil Productivity
It is an index of available nutrients to plants.	It is a broader term used to indicate yields.
It is one of the factors for crop production. The other factors are water supply, slope of the land, depth of water tableetc.	It is the interaction of all the factors that determine the magnitude of yields.
It can be analyzed in the laboratory.	It can be assessed in the field under particular climatic conditions.
It is the potential status of the soil to produce crops.	It is the resultant of various factors. influencing soil management.

All the productive soils are fertile, but not all fertile soils are productive.

Factors affecting soil productivity

The factors affecting soil productivity include all those which affect the physical, chemical and biological conditions of the soil environment in which plants grow. They include all the practices that affect fertility, the water and air relationships and the activity of the biological agents such as insects, pests, diseases and microorganisms.

I. **Internal factors**: may be called as genetic or hereditary factors which cannot be manipulated such as soil type, texture etc.

II. **External factors:** may be regulated to certain extent, They include

a. Climatic factors: like precipitation (rain fall), solar radiation, atmospheric gases (CO_2, NO_2, N_2O, O_2), wind velocity etc.

b. Edaphic or Soil factors: Soil moisture, soil air, soil temperature, soil mineral matter, inorganic and organic components, microorganisms, soil reaction.

c. Biotic factors:

   i. Plants: have competitive and complementary nature, competition between weeds and crop plants, plants growing as parasites.

   ii. Bacteria of symbionts, free living.

d. Animals: earth worms, small and large animals.

e. Physiographic factors: geological strata (parent materials), topography (altitude, steepness of slope).

f. Anthropogenic factors: human factors including skill and efficiency of cultivation by man.

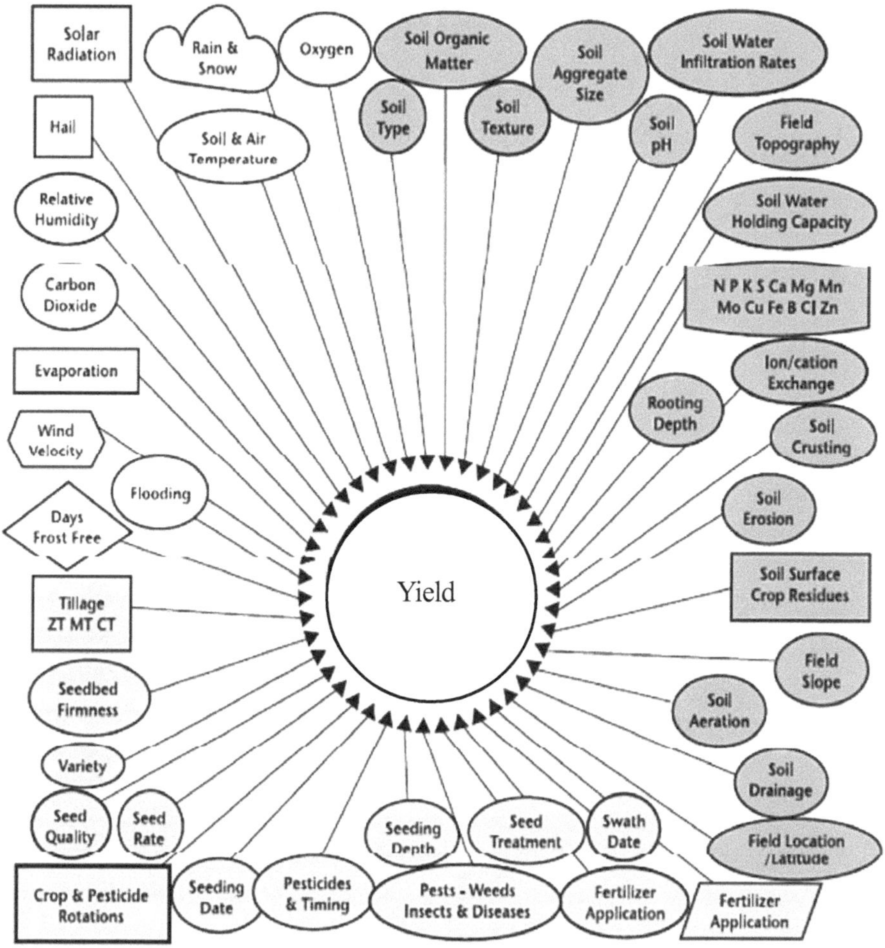

Fig. 1: Crop production factors

Chapter 2

Essential Plant Nutrients

Terminology

- *Nutrient:* Nutrients are substances required by an organism for their normal growth and reproduction.
- *Plant Nutrient:* The plant nutrient is a "food" which is composed of certain chemical elements often referred to as 'plant nutrient' or plant food elements considered very essential for growth and development of plants.
- *Nutrition:* The supply and absorption of chemical compounds needed for growth and metabolism of an organism.

Plant nutrient elements are broadly grouped into two types.

- Essential Nutrients/ Elements
- Beneficial Nutrients/Elements

A. Essential nutrients/elements

Plants absorb or utilize more than 90 nutrient elements from the soil and other sources during their growth and development and about 64 nutrients have been identified in plants by their tissue analysis.

But all are not essential for their growth and development. They require only 17 elements/nutrients. These 17 have been recognized as essential elements. They are;

Carbon (C), Hydrogen (H), Calcium (Ca), Magnesium (Mg), Sulphur (S), Iron (Fe), Manganese (Mn), Zinc(Zn), Copper (Cu), Boron (B), Molybdenum (Mo), Chlorine (Cl), Nickel (Ni), Oxygen (O), Nitrogen (N), Phosphorous (P), Potassium (K).

Of these element C, H, O together constitute 95-96% (C-45%, O-45%, H-6%). Subsequently N, P and K constitute 2.7% in plants. The other elements constitute only 1.3-1.4%. But all have definite roles to play in the growth and development. Among these Nickel is the latest nutrient addition to the list in 1987.

Classification of essential nutrients

Essential nutrients are classified in to two major groups based on relative utilization or absorption by the plants and also based on their biochemical behavior and physiological functions.

Based on relative utilization or absorption by the plants

A. Macro or Major Nutrients

B. Micro Nutrients

Further macronutrients are classified into two types

Primary Nutrient: Nitrogen, Phosphorus and Potassium. These three elements are also called as fertilizer elements.

Secondary Nutrients: Calcium, Magnesium and Sulphur.

Table: Classification based on their biochemical behaviour and physiological functions.

Group	Nutrient elements	Biochemical Functions
Group-I	C,H,O, Ca	They are basic structural elements. They are Major constituent of plants (carbohydrates, proteins and fats) and organic matter. These elements are also involved in enzymes process. They provide energy for growth and development by oxidative break down.
Group-II	N,P & S	Accessory structural elements of the more active and vital living tissues. Essential component of metabolically active compounds like amino acids, proteins, enzymes and non-proteinaceous compounds. They involve in energy storage (ATP & ADP) and transfer (Phosphate esters).
Group-III	K, Ca, Mg	Regulators & carriers for the most part of plant metabolism. They involve in synthesis and translocation of carbohydrates, maintain ionic charge balance and induce enzyme activation.
Group-IV	Fe, Mn, Zn, Cu, B, Mo, Cl.	Catalysts and activators. These elements Involve in oxidation-reduction reactions, chlorophyll synthesis and also exists in organic combinations.

B. Beneficial nutrients/elements

Beneficial elements are the mineral elements which stimulate the growth and have beneficial effects even at very low concentration.

They are not essential or essential only for certain plant species under specific conditions. They are also known as 'potential micro-nutrients'.

These elements have been found to affect the uptake, translocation and utilization of other essential elements, help in production of essential metabolite by activating enzymatic system/action and also counteract the toxic effects of some other elements or anti metabolites.

Eg: Silicon (Si) for rice, Sodium (Na), Aluminum (Al), Cobalt (Co), Selenium (Se), Iodine (I), Gallium (Ga) and Vanadium (Va).

Definitions

- **Macro or Major Nutrients:** They are the nutrients utilized by the plants in relatively large amounts (quantity) for their growth and development.

  Eg: C, H. O. N, P, K, Ca, Mg and S (C, H and O are abundantly present in the atmosphere and need not be applied through fertilizers).
- **Primary nutrients:** are those nutrients required relatively in large quantities by the plants for their growth and development. These are also designated as 'fertilizer elements' because, deficiency of these elements is corrected by application through fertilizers. *Eg*: N, P and K
- **Secondary nutrients:** are those nutrients which are required by plants in moderate amounts. They are called secondary because they are unknowingly supplied through fertilizers and other amendments. However their role in nutrition is not secondary but they are given secondary importance in its supply and management. *Eg*: Ca, Mg & S Ex: When SSP is applied as a fertilizer for P it supply Ca and S . Dolomite applied as a liming material supply Ca and Mg.

Ammonium Sulphate added as N fertilizer will supply S

Micronutrients: The nutrients which are required by plants in relatively smaller quantities for their growth and development, but these are equally important and essential to plants as macronutrients. They are also called as trace/rare/nano elements. These include Fe, Mn, Zn, Cu, B, Mo, Cl and Ni

Criteria of essentiality of nutrients

This concept was propounded by Arnon and Stout (1939) and they considered 16 elements essential for plant nutrition. For an element be regarded as an essential nutrient, it must satisfy the following criteria;

A deficiency of an essential nutrient element makes it impossible for the plant to complete the vegetative or reproductive stage of its life cycle.

The deficiency of an element is very specific to the element in question and deficiency can be corrected /prevented only by supplying that particular element.

The element must directly be involved in the nutrition and metabolism of the plant and have a direct influence on plant apart from its possible effects in correcting some micro-biological or chemical conditions of the soil or other culture medium.

Functions and deficiency symptoms of essential nutrient elements-

C,H,O,N, P and S are elements of life as they form covalent bonds with C, thus entering into organic combination. When plants lack essential nutrients or elements they will not look as they are, but look unhealthy and may even die. The symptom of a nutrient deficiency can range from yellowing and poor growth to flower and fruit failure.

Terminology

- **Deficient:** When an essential element is at a low concentration in plant that severely limits the plant growth and produces more or less distinct deficiency symptoms on plants. Under such conditions the yield of crop will be low / the quality of produce will be inferior.
- **Insufficient:** When the level of an essential nutrient is below their actual content in plant or available in an inadequate amounts that also affect the plant growth and development.
- **Toxic:** When the concentration of an element in plants is very high this affects the plant growth severely and produces toxicity symptoms on plants.
- **Excessive:** When the concentration of an essential nutrient is sufficiently high but not toxic. It results in a corresponding shortage of other nutrients.

Functions and deficiency symptoms of essential nutrients

A. Carbon, Hydrogen & Oxygen

- Carbon, Hydrogen and Oxygen form about 95% of the dry weight of plants and are obtained from CO_2 and H_2O. They are converted in to simple carbohydrates by photosynthesis and ultimately elaborated into complex amino acids, proteins and protoplasm. These are the major components of carbohydrates, proteins and fats.

Functions

- They play a dominant role in the process of photosynthesis and respiration in plants.
- They are involved in the formation of simple as well as complex organic compounds like carbohydrates, starch proteins *etc.*
- Maintaining the structure of the plant cells.
- They provide 'energy' required for the growth and development of plant by oxidative break down of carbohydrates, proteins and fats during their cellular respiration

Primary nutrients

N, P and K are considered as primary nutrients.

1. Functions and deficiency symptoms of Nitrogen

Nitrogen plays a key role in the nutrition of plants. It is one of the principal growth promoting nutrient elements. Green plants are more markedly influenced by the deficiency of nitrogen than by any other element. It is absorbed by plants in the ionic form of NO_3^-, by most of the plants. Some plants require NH_4^+ form (rice). When applied as foliar nutrition, NH_2 (amide from) is also absorbed. It has got most recognized role in the plant metabolism as it performs the following vital functions.

Functions of Nitrogen in plants

The Nitrogen is mainly involved in Photosynthesis of plants as it isan essential constituent of chlorophyll.

It is very basic constituent of plant life, because, it forms essential constituent of proteins, nucleotides, phosphatides, alkaloids, enzymes, hormones, vitamins etc.,

It promotes better Vegetative growth and adequate supply of nitrogen promotes rapid early growth and imparts dark green colour to plants, improves quality and succulence of leafy vegetables and fodder crops.

- It stimulates the formation of fruit buds; increases fruit set, and improve quality of fruits.
- It governs the better utilization of Potassium, Phosphorus & other elements.

Deficiency symptoms of Nitrogen in plants

Nitrogen is highly mobile element in plants and so deficiency is exhibited in older/ bottom leaves. The striking deficiency symptoms are Yellowing of older leaves due to inhibition of chloroplasts and chlorophyll synthesis. As the deficiency of Nitrogen becomes severe "Chlorosis" of leaves is observed. Lower leaves pale green or bluish then turn yellow (chlorosis); leaves drop, the oldest leaves fall first; leaves are small; stems thin; plant lacks vigor; growth is spindly or stunted.

- Plants become dwarfed or stunted growth.
- Tends to advance the time of flower bud formation and reduce yield.
- Fruits become hard, small, low bearing capacity of trees,
- Reduces fertilization, premature dropping and fruits may become seedless.
- Severe deficiency leads to Necrosis of plant leaves (complete death of leaf)

Management of deficiency

Source of N: Manure, bone meal, blood meal (dried blood), fish meal, fish emulsion (also contain phosphorus and potassium, in small amounts), cotton seed meal (also contains small amount of phosphorus and even smaller amount of potassium), coffee grounds (also contains very small amounts of phosphorus and potassium), soybean meal (also contains small amount of potassium and even smaller amount of phosphorus), composted legumes (peas, beans, peanuts), ammonium sulphate or nitrate.

Excessive N in plants leads to more vegetative growth. Leaves become Leaves dark green, more succulent and more susceptible to pest and disease attack. Lodging of plants may occur, excessive leaf growth at the expense of buds and fruits. Reduces the sugar content of plant, storage and keeping quality of fruits or leaves and prolong the growing period and delay the reproductive phase of plant and crop maturity.

Nitrogen deficiency symptoms on plants

Fig 2.1: Bitter gourd (Stunted plant growth; Old leaves turn yellow in colour.)

Fig. 2.2: OnionLeaves become yellowish green erect and upright curled, wilted and dwarf. At maturity tissue above bulbs become soft.

Functions and deficiency symptoms of Phosphorus

Phosphorous is a constituent of essential cell components such as phytins, phosphoproteins, phospholipids, nucleic acids (DNA, RNA), co-enzymes (NAD & NADP), ATP and other high energy compounds. It is also a structural component of cell membrane, chloroplasts, mitochondria and meristematic tissues. Plants absorb the Phosphorus as $H_2PO_4^-$ and HPO_4^{2-} ionic form. Phosphate compounds act as "energy currency" within plants. It is highly mobile in plants but immobile in soils.

Functions

- Involved in Energy storage and transfer. Also carry various metabolic processes in plants.
- Involved in cell division and development of meristematic tissue and thus it improves better vegetative growth of plants.
- Important for root development and stimulates root growth.
- Helps in primordial development, flowering, seed formation, ripening of fruits germination of seeds and also early maturity of crops.
- It is essential for formation of starch, proteins, nucleic acids, photosynthesis, nitrogenmetabolism, carbohydrate metabolism, glycolysis, respiration and fatty acid synthesis.

Deficiency symptoms of Phosphorus in plants

Fig. 2.3: Onion Slow growth, maturity blazed. Leaf colour becomes light green and bulbs have few dried outer peals. Tip burn in older leaves.

- Stunted and slow growth of plants due to its effects on cell division and meristematic tissue development.
- Leaves are small and defoliation starts from the older leaves and premature leaf fall young leaves look pale.
- Purplish discoloration of foliage due to anthocyanin pigment.
- Plants develop dead necrotic areas on the leaves, petioles or fruits.
- Slender and woody stem with under developed roots are characteristics symptoms.
- Delay in flowering and ripening of fruits, inferior quality, shedding of blossom, inflorescence becomes small and premature fruit falling.
- Inhibit the sugar synthesis or abnormally high sugar levels in plant.

Sign of excess: Essential elements may be tied up.

Functions and deficiency symptoms of Potassium

• Potassium is indispensable in the plant nutrition and needs to be supplied in relatively large quantities to fruit crops and field crops. Plants absorb K from the soil as K^+ ion and it is mobile in nature in plants. Potassium does not enter in to the composition of any of the constituents of the plant cells such as proteins, chlorophyll, fats and carbohydrates. It primarily occurs as soluble inorganic salts and occasionally as salts of organic acids. It is abundant cation in the cytoplasm, meristematic regions, cell sap. It is considered as Quality element for many crops.

Functions

- Potassium is responsible for osmoregulation and controls cell turger pressure.
- It has important role in pH stabilization, enzyme activation, protein synthesis, stomata movement (closing and opening), cell extension and photosynthesis.

Impart drought/heat/frost resistance to plants as it regulates transpiration and water conditions in the plant cell. It improves water use efficiency

- Impart pest and disease resistance to plants
- Required for ATP synthesis and better N use efficiency by favoring the protein formation.

- Plants become strong and stiff; thus it reduces lodging of plants.

Essential in the formation and transfer of starch and sugars especially in potato, sweet potato, turnip, banana, tapioca.

Nutrient ratios (balanced nutrition) of either K/Ca or K/Mg are important in crops.

It also plays a key role in enzyme activation. It is involved in the functions and activities of more than 60 enzymes.

If N increases yield, K imparts stability

Deficiency symptoms

- Weakening of stem and Lodging of crops and easy susceptibility to pest and diseases.
- Scorching of leaves and burning appearance of leaf margins and tip Poor keeping quality of fruits. The quality of fruits and vegetables decreased.
- Marginal necrosis and burning of leaf tips.
- Stunted growth, shortening of internodes.
- It causes great disturbance in the water economy of plants (K/Cainteraction) and more water is lost per unit dry matter.
- Poor sprouting of vines.
- Severe attack of the grapes with *Botrytis cinerea* due to K deficiency.

Potassium deficiency symptoms

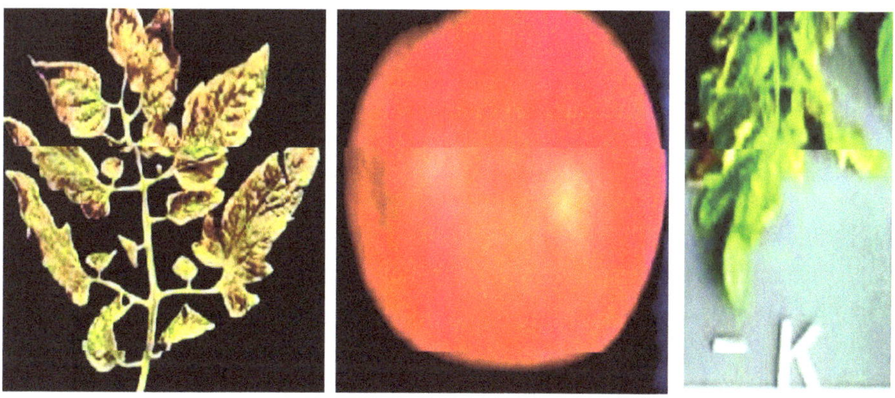

Fig. 2.4: K deficiency in cucumber

- Leaflets slight marginal and interveinal chlorosis followed by brown marginal scorching;
- **Sign of deficiency:** Lower leaves spotted, mottled or curled; leaf tips and edges turn yellow and bronze, then brown and appear dry and scorched margins curl forward; plant vigour reduced; plant susceptible to wilting and wilt diseases; fruit is small; fruit is thin skinned; fruit colour is poor; flavour is poor.
- **Sign of excess:** Fruit coarse and poorly coloured fruit; reduced absorption of magnesium and calcium.

Secondary nutrients

Ca, Mg, and S are called as secondary nutrients.

1. Calcium (Ca)

It is immobile in plants and exists as deposits of calcium oxalate, calcium pectate in the middle lamella of cell wall and $CaCO_3$ and $CaPO_4$ in cell vacuoles. Although calcium is present in plants in relatively higher proportion as compared with other elements, its actual requirement by plants is not much higher than that of a primary nutrient.

Functions

1. It is a constituent of the cell wall andpromotes early root development maintains the permeability ofmembranes
2. It is required for cell divisions and chromosome stability, cell wall construction, cell elongation of the shoot and root.
3. Stabilizing the pectin of the middle lamella in the cell wall by forming calcium pectate.

Thus Ca brings resistance against diseases.

4. Effect on fruit quality and increases in the firmness of the fruit.
5. Indirectly influences many enzyme systems and maintain cation-anion balance (by acting as a counter ion).
6. Calcium is also responsible for enhancing pollen germination and growth.

Calcium is required for avoiding the toxicity effects of heavy metals inplants.

Deficiency symptoms

- Deficiency is first observed on the young leaves and growing tips (immobile in plants).

- Leaves become small, distorted, cup shaped, crinkled and malformation of leaves (It resembles boron deficiencies)
- Terminal buds may deteriorate and die back in fruits trees. Root growth is impaired.
- Destruction of cell well structure results in disturbance of nuclear and cell division.
- Fruit quality is reduced, loss of fruit fleshy, blossom-end of fruit rots; cavities in tomatoes; black heart; black roots. Sometimes rotting of fruits and susceptible to fungal disease.

Sign of excess: Intake of potassium and magnesium is reduced.

Calcium deficiency (blossom end rot) on a tomato

Fig. 2.5: Ca deficiency of cucumber

Calcium deficiency

Death of growing point and die-back of main stem from tip; die-back of leaves, progressing from terminal leaflets and of flower and fruiting trusses.

Calcium deficiency: Dying off of terminal leaflets and flowers; leaves purplish brown tinting

Ca deficiency in coffee

Ca Deficient root Healthy root (coffee)

(Leaves bronzed along edges, cupped downward; new leaves dead; eventual dieback of shoot tips)

2. Magnesium (Mg)

Mg is a constituent of the chlorophyll molecule and located at its centre, without which photosynthesis by plants would not occur. It is a mobile element and plant absorb as Mg^{2+} ionic form.

Functions

- Very much essential for photosynthesis.
- It is involved in the regulation of cellular pH, cation-anion balance and turgur regulation of cells.
- Necessary for protein synthesis.
- Activator of enzymes in carbohydrate and ATP metabolism.
- Essential for the formation of oils and fats
- It is required for stabilization of cell membranes.

Deficiency

- Interveinalchlorosis of lower leaves and in extreme cases becomes necrotic.
- Leaves remain in small and brittle even in final stages.
- Twigs may become weak and premature dropping of leaves results in heavy loss of fruit crops.
- Inhibits nitrate reduction and the production of photo hormones.
- Stalk necrosis or stem 'Die back' in a Vine plant.

Excess of Mg absorption becomes poisonous leads to browning of roots as a result growth is ceased and death of roots and leaves. It can be counteracted by CO_2 antagonistic action. Absorption of calcium and potassium is reduced.

Fig. 2.6: Mg deficiency in cucumber (Central interveinal chlorosis and green marginal bands)

3. Sulphur

It is abundant in plant, particularly in the leaves. Plant absorbs as sulphate (SO_4^{2-}) form.

It is not easily translocated in plants.

Functions

1. Required for synthesis of the S-containing amino acids like cystine, cysteine and methionine, which are important for protein synthesis.
2. Role in photosynthesis by involving in structural formation of chlorophyll in leaves.
3. It is a constituent of proteins and volatile compounds responsible for the characteristic taste & smell of plants in the mustard and onion families.
4. It enhances oil synthesis in crops
5. It is a vital part of Ferrodoxins (Non Heme iron, sulfur protein), S- adenosyl methionine.

Deficiency symptoms

1. Pale yellow or light green leaves in younger leaves (Deficiencies resemble those of nitrogen)
2. Stalks are short & slender, growth is retarded.
3. Fruits often do not mature fully & remain light green in colour.
4. In Brassica species, leaves shows cupping & curling.
5. Cell division is retarded & fruit development is suppressed.

6. Disrupts N metabolism, reduces protein quality & induces starch (carbohydrate) accumulation.

Fig. 2.7 : S deficiency in onion (Yellowing of young leaves)

S-Toxicity: Sulphide injury, necrosis of the leaves. (Sulphur burn from too low pH).

Micronutrients

These are essential plant nutrients required in minute quantities. There are 7 micronutrients namely Iron, Manganese, Boron, Molybdenum, Copper, Zinc and Chlorine. Micronutrients are also called minor elements or trace elements.

1. **Iron: (Fe):** It is the first micronutrient to be discovered as an essential element for plant life. Iron present in chloroplasts as a "ferrodoxin" compound. Plants obtain as Fe^{2+} and Fe^{3+} forms and also as chelated Fe form. Immobile element within the plant; as such iron deficiency is noticeable in younger leaves at the growing region.

Functions

- Involved in biosynthesis of chlorophyll and in the synthesis of chloroplast proteins
- Activates several enzymes involved in respiration.
- It brings about oxidation-reduction reactions in the plant.
- It regulates respiration, photosynthesis, reduction of nitrates and sulphates.

Deficiency symptoms

1. Interveinalchlorosis of younger leaves and generally called as "Iron chlorosis" or lime induced chlorosis. On severe deficiency leaves become "Pale white".
2. Reddish-brown necrotic spots along the leaf margins of young shoots in tree crops.
3. In Brassica necrotic terminal buds at early seedling stage.

Fig. 2.8 : Fe deficiency in cucumber (Interveinal chlorosis of the youngest leaves ends up in a totally bleached leaf)

Manganese (Mn)

It is absorbed by plants as Mn^{2+}form from the soil. It is translocated to the different plant parts where it is most needed.

Lime inducedchlorosis-Fe deficiency
Grape orchard

Functions

- Involved in oxidation-reduction reactions and electron transport in photosystem II
- It is directly or indirectly involved in chloroplast formation and their multiplication.
- It activates large number of enzymes and acts as a co-factor and catalyses most of the enzymes
- It helps in movement of Iron.

Deficiency symptoms

- Interveinal chlorosis on old leaves similar to Iron chlorosis.
- Speckled yellow of sugarbeet-leaves develop interveinal yellowish green chlorotic mottling and leaf margins role upwards.
- Depresses inflorescence and fructification and results in stunted leaf and root development

Fig 2.9: Mn deficiency in cucumber (Interveinal chlorotic mottling distributed over entire surface of leaf, mottled areas become necrotic) (See colour version on page 141)

Manganese toxicity (soil acidity complex): Tomato shoot; Stem and petioles, especially nodal areas, necrotic lesions; leaves wither and hang down.

Copper (Cu)

Minute quantities of copper are necessary for normal growth of plants. Copper salts are poisonous even in exceedingly small concentrations. It is absorbed as cupric ion (Cu^{2+}). Its function is almost similar to those of Fe. It is immobile element in plants.

Functions

- It acts as electron carriers in enzymes which bring about oxidation-reduction reaction in plants.
- Helps in utilization of iron in chlorophyll synthesis.

- Influence on cell wall permeability and nitrate reduction.
- Play a role in the biosynthesis & activity of ethylene in ripening fruit.
- Promote the formation of vitamin-A in plants.
- Influence on pollen formation & fertilization.

Deficiency

- Narrow, twisted leaves and pale white tips, interveinal chlorosis mottling of leaves.
- In fruit trees "die-back" (terminal bud wither and die) is most common.
- It affects fruit formation much more than vegetative growth.
- The critical stage of Cu deficiency induces pollen sterility in microsporogenesis. Reduced fruit set and number of flowers.

Fig. 2.10: Cu deficiency in onion-Scales become this and pale yellow colour. The bulbs lack firmnesss and solidity and early maturation leaves slow tip burn with Chlorate mottling, permanent loss of turger in young leaves. (See colour version on page 142)

Boron (B)

Boron is present especially at the growing points and in the conducting tissue. This element being a non metal doesn't appear to be a part of any enzyme system. Plants absorb B as $H_3BO_3^-$, $B_4O_7^{2-}$, H_2BO^{3-}, and HBO_2^{-3}& BO_3^{2-}. It is immobile element in plants.

Function

- Essential for cell division in the meristematic tissues. Maintains cell turgidity.
- Involved in proper pollination, pollen formation, pollen tube growth/ flowering and fruit or seed set.

- Important role in the fertilizing process of plants and during blossom period its requirement is high.
- It influences carbohydrates and N-metabolism and also Ca.
- Translocation of sugars through cellular membranes and prevents the polymerization of sugars.
- It enhances rooting of cutting through oxidation process.
- It has role in hormone movement and action.
- It gives resistance for pest and disease infection, *e.g.*: virus, fungi & insects. Role in water relations *i.e.,* prevents hydration of root tips & thus strengthens the plant roots
- Acts as a regulator of potassium/calcium ratio in the plant. Solubility & mobility of Ca increases.

Deficiencies

- Young leaves may be deformed, appear like a "rosette", cracking and cork formation in stems, stalks and fruits, thickening of stems and leaves, reduced buds, flowers and seed production.
- Premature seed or fruit drop.
- 'Hen and Chicken disease' in grapes bunches i.e. fruits of vine with small & long berries.
- Deformed fruits of papaya tree.
- Vine plant with thickened internodes. Poor fructification and development of the berries.
- In mango, leaves become pale green distorted &brittle .
- Browning or hollow stem of cauliflower.
- Heart rot disease' in fruits of the sweet melon (*Cucumismelo*), sugar beet & marigold.
- Interruption in cell wall formation and differentiation and then necrosis.
- Flowers wilt, die and persist on the tree. This phenomenon is called "Blossom Blast".
- Tissue break down and preventing sugar and starch accumulation in the leaves.
- Excessive formation and accumulation of phenolics.
- Bitter orange fruits with thickened peels or rinds & blackish discoloration.

Fig. 2.11: Distortion of newer leaves (in severe cases the growing point dies) and the appearance of a broad yellow border at the margins of the oldest leaves. Young fruit can die or abort; abortion rates are high. Stunted development and mottled yellow longitudinal streaks, which develop into corky marking (scurfing) along the skin.

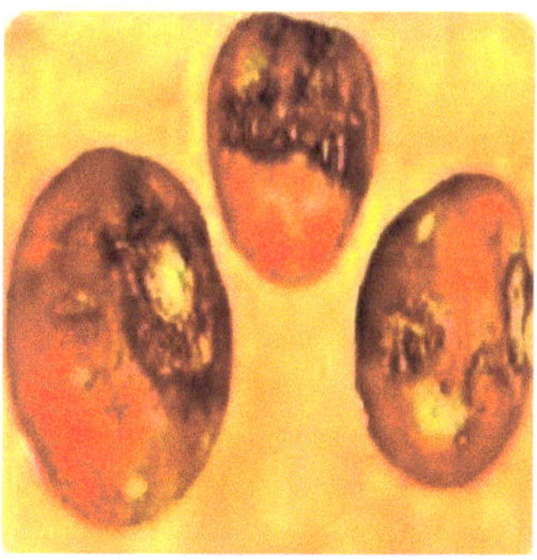

Boron deficiency
Fruits pitted and corky areas in skin; ripening uneven (calcium deficiency also).

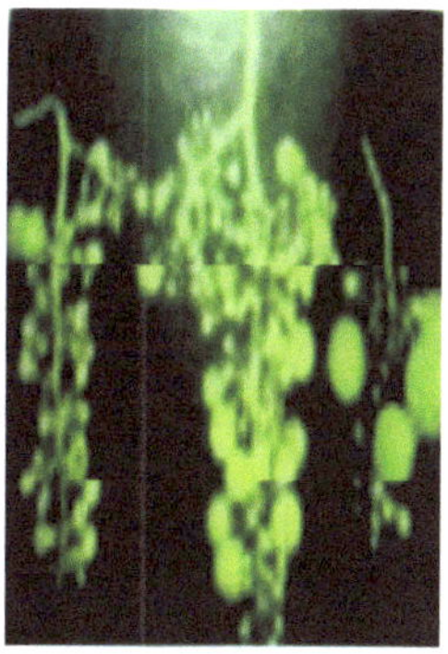

Hen and Chicken
Disease in grapes bunches i.e. fruits of vine with small & long berries

Deformed fruits of papaya tree due to B deficiency

B Toxicity

Yellowing of the leaf tip and leaf margin which spreads towards the midrib leaves become scorched and may drop early.

Molybdenum (Mo)

Required by plants in small quantity, plant absorb as MoO_4^{2-} form. It is structural components of Nitrogenase enzyme and constituent of nitrate reductase.

Functions

- Essential role in iron absorption and translocation in plants, protein synthesis and N- Fixation in legumes.

- Brings oxidation and reduction reactions especially in the reduction of NO_3 to NH_4.
- It acts as a bridge or link in transferring electrons.
- Role in phosphate system and ascorbic acid synthesis.

Deficiency

- Reddish or purplish discoloration of leaves, chlorosis and marginal necrosis of leaves.
- Marginal scorching and rolling or cupping of leaves, "Yellow spot" disease of citrus and "Whiptail" in cauliflower is commonly associated.
- NO_3 accumulation in plants thus inhibits the utilization of N for protein synthesis.

Scald- The tomato leaves show chlorotic mottling and inward curling of leaf margins. The older leaves of radish show interveinal yellowish green mottling, they become papery, bleached and have downward cupping. Similar symptoms appear in beans, which are called as 'Scald'.

Yellow spot of citrus: In cereals the middle leaves becomes golden yellow at the tip during ear head emergence stage. Affected leaves become dry and papery while younger leaves become spirally twisted. Similar symptoms appear in citrus and the deficiency disease is called as "yellow spot of citrus".

Molybdenum deficiency
Left: Healthy leaf receiving molybdenum. *Right:* Molybdenum deficient leaf, leaflets, incurred margins. Intervene! chlorate mutt and death of tips

Molybdenum deficiency
Leaflets somewhat chlorotic, strongly incurred and die back from tips.

(See colour version on page 142)

Zinc (Zn)

Zinc has limited mobility in plants and immobile in soil and plant absorb as Zn^{2+} form.

Functions

- Zn is a constituent of several enzymes systems which regulate various metabolic reactions in the plant.
- Influences the formation of some growth hormones in the plant like IAA, and Auxin.
- Helpful in reproduction of certain plants.
- Role in photosynthesis and involved in chlorophyll synthesis, protein synthesis.
- Involved in alcohol dehydrogenase activity in fruit trees.

Deficiency

- Chlorotic and Brown rusty spots on leaves.
- Lower Auxin level.
- Drastic decrease in leaf area and leaf deformation (Rosetting), stunted growth (shortage of internodes).
- Under severe deficiency the shoot apices die (dieback) and diffusive or mottled leaf
- The rate of protein synthesis is drastically reduced and amino acids and amides accumulate.

Leaf shows an advanced case of interregional necrosis. In the early stages of zinc deficiency the younger leaves become yellow and pitting develops in the interregional upper surfaces of the mature leaves.

A Zinc deficiency can cause stunting, with white spotting between veins, in tomatoes

Chlorine (Cl)

Chlorine is readily taken up by plants and its mobility in short and long distance transport is high. It does not form constituents of organic substance but act only in ionic form. The plant requirement for chlorine is rather quite high as compared to other micronutrients. The exact role of Cl in plant metabolism is still obscure.

Functions

- Involved in the evolution of "Oxygen" by chloroplasts in photo system-II.
- Associated with turgor production in the guard cells by the osmotic pressure exerted by K^+ ions
- Role in stomata regulation (opening & closing).
- Water splitting in photo system-II.
- Act as a bridging ligand for stabilization of the oxidized state of Mn.

Deficiency

- Chlorosis and burning of tips and margin of leaves. In tomato, leaves become chlorotic and later bronzed.

Over wilting effect and leaf fall, yielding ability decreases.

Diagramatic representation of common important nutrient deficiencies of plants

Chloride toxicity on many crops- Bronze or yellow colors of leaves with brown or scorched leaf margins

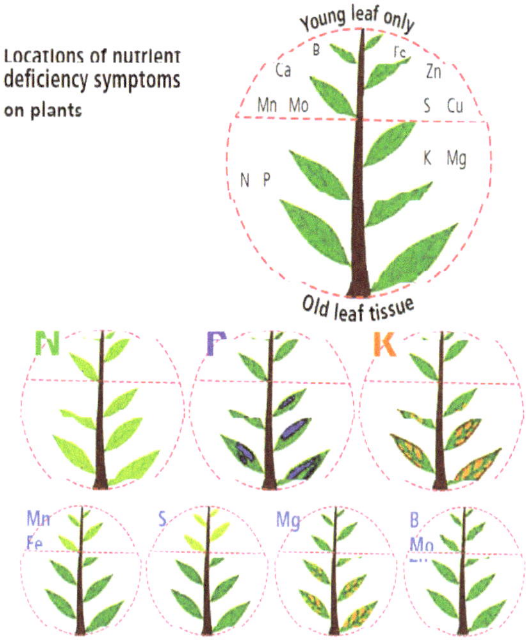

Nutrient	Location of	Chlorosis? symptoms	Leaf margin	Leaf color, shape necrosis?
N	All leaves	Yes	No	Yellowing of leaves, leaf veins
P	Older leaves	No	No	Purplish patches
K	Older leaves	Yes	Yes	Yellow patches
Mg	Older leaves	Yes	No	Yellow patches (in oil palm) or interveinal chlorosis (in rice and maize)
S	Young leaves	Yes	No	Yellow patches
Mn, Fe	Young leaves	Yes	No	Interveinal chlorosis
B, Zn, Mo	Young leaves	-	-	Deformed leaves

Diagramatic representation of common important nutrient deficiencies of plants (See colour version on page 143)

Functions of Essential nutrient elements is given in the table

Nutrient	Function
Carbon	It is the basic molecular component of carbohydrates, proteins, lipids and nucleic acids.
Oxygen	It occurs in all the organic compounds of living organisms.
Hydrogen	This element plays a central role in plant metabolism. It is very important in ionic balance, as the main reducing agent, and plays a key role in energy relations of cells.
Nitrogen	It plays a significant role in the synthesis of important organic compounds, amino acids, proteins, nucleic acids (RNA, DNA), enzymes etc.

Phosphorus	It is an important component of proteins and enzymes, nucleic acids (DNA and RNA) and phytin. Phosphorus is also involved in various energy transfer reactions of adenosine triphosphate and diphosphate (ATP and ADP).
Potassium	It helps in osmotic and ionic regulation. Potassium is a cofactor or activator for many enzymes of carbohydrate and protein metabolism.
Calcium	It is involved in cell division and plays a major role in the maintenance of membrane integrity.
Magnesium	It is a component of chlorophyll and a cofactor for many enzymatic reactions.
Sulfur	Somewhat like phosphorus, it is involved in plant cell energetics. It plays an important role in plant lipid synthesis.
Iron	It is an essential component of many heme nonheme iron enzymes and carriers, including the cytochromes (respiratory election carriers) and the ferredoxins. The latter are involved in key metabolic functions such as nitrogen fixation, photosynthesis and electron transfer.
Zinc	It is an essential component of several enzyme systems (dehydrogenases, proteinases and peptidases including carbonic anhydrase, alcohol dehydrogenase and others).
Manganese	It is involved in the 02-evolving system of photosynthesis and is a component of the enzymes arginase and *phosphotransferase*.
Copper	It is a constituent of a number of important enzymes, including cyctochrome oxidize. Ascorbic acid oxidize and lactase.
Boron	The specific biochemical function of boron is unknown, but it may be involved in carbohydrate metabolism and the synthesis of cell wall components.
Molybdenum	It is required for the normal assimilation of nitrogen in plants. It is an essential component on nitrate reeducates as well as nitrogen (N_2 fixation enzyme).
Vanadium	It stimulates biological nitrogen fixation of bacteria (*Azotobacter*, *Rhizobium* sp.) similarly to Molybdenum. Vanadium has an important role in the food chain (essential for animals and humans)
Chlorine	It is essential for photosynthesis and as an activator of enzymes involved in splitting water. It also functions in osmoregulation of plants growing on saline soils.

Chapter 3

Relationship Between Plant Nutrient Concentration and Plant Growth/Yield Luxury Consumption of Nutrients and Nutrient-Interactions in Plants and Soils

The following graph is a visual representation of how plant growth and/or yield is affected by nutrient concentrations

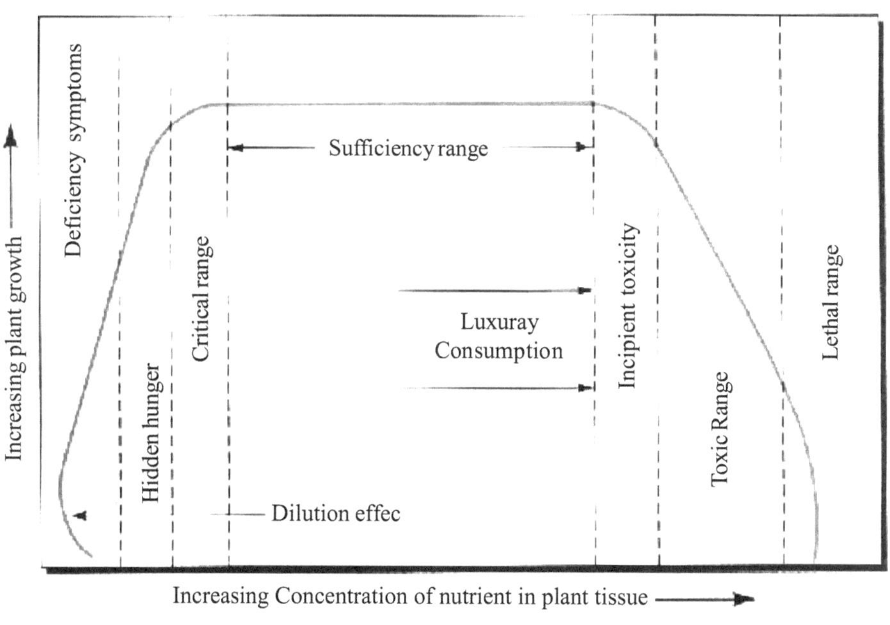

The Steenberg effect: Known under extreme deficiency, rapid yield increase can cause some decreases in nutrient concentration

Luxury consumption

It is the tendency of some crops to absorb and accumulate nutrients far in excess of their actual needs if it is present in sufficiently large quantities in the soil. Potassium is one of the nutrient elements which is subjected to luxury consumption. The absorption pattern of different nutrients by plants varies greatly among the plant species and with their age and growth stages.

Consumption of nitrogen by plants

Plants absorb N mostly in nitrate (NO_3^-) form or in ammonical (NH_4^+) form by some plants. Plants usually absorb the N more during active growing period, but therate varies with time. The amount of nitrogen absorbed is at maximum when the plants are young and gradually declines as the plants age. Plants can absorb extra nitrogen when it is available and store it to be used later if needed. An over supply of N generally produces dark green, succulent, vegetative growth. In such cases a decline in seed production of grain crops, fruit production in tomatoes and some tree crops occurs. In sugar beets, sugar content decreases and in and potatoes, tubers become watery. The negative effects of too much of N on growing plants can be lessened if the P and K supplies are adequate.

The average utilization of applied N by crops is around 50 percent but with proper nitrogen management strategies the efficiency as high as 80 % or more can be increased. Low N use efficiency may be attributed to various losses such as volatilization of ammonia in alkaline soil, Denitrification of Nitrate ions in flodded soil, Leaching loss of nitrates in coarse textured soil, soil erosion/run off and ammonium fixation in clay lattices.

Consumption of phosphorus by plants

Phosphorus application, unlike N is known to benefit the growth and productivity of more than one crop in rotation. The residual P contributes more of P to crop nutrition. Responses to applied P depend on soil properties, initial available P, variety, level of N applied and management practices.

Phosphorus is absorbed as phosphate ions such as $H_2PO_4^-$ and HPO_4^{2-} form. It is concentrated more in the reproductive parts of plant and in seeds. Harvested crops contain considerable amounts of P. In general, seed crops contain largest percentages of P, and forage crops contain moderate percentages.

Consumption of 'P' by the crops is very less after their application to soil and it accounts even less than 10 % and remaining amount will be useful later. This is mainly because; P is subjected to immobilization or fixation (retention/adsorption/precipitation/sorption) and undergoes various transformations which render it unavailable to plants. P fertilizers are not easily and completely soluble in water and their mobility is less within the soil. Therefore in order to get maximum benefit from them we have to adopt suitable methods and time of application.

Consumption of potassium by plants

Potassium uptake is often equal to or more than that of nitrogen. It is absorbed by plants in K^+ form. Crop species differ in their K requirement. Tuber crops like potato, vegetables like cauliflower and cabbage, forages like alfalfa and fruits like banana, grapes and pineapple, plantation crops like coconut, tea, rubber are among the heavy feeders of K. High crop yields and higher rates of N and P application accelerate K uptake from the soil. Crop responses to K are large on laterites, red and yellow and mixed red and black soils. Plants absorb and accumulate K far in excess of their needs if it present sufficiently in soil without affecting the metabolic activity or without any plant response. This is called as Luxury consumption.

Potassium is also subjected for various losses

1) Leaching losses of K- Especially in sandy soils and soils rich in kaolinite located heavy rainfall area.
2) Soil erosion losses- It also leads to considerable loss of total K from the soil.
3) Fixation of K by clay complex of illite type.

Consumption of secondary nutrients by plants

The amount of secondary nutrients removed by crops depends on the soil type, crop species, fertilizer sources and yield level. Generally, legumes and root crops remove more Ca and Mg than do cereals and other grasses. Cereals may remove 10-20 kg Ca per ha, a good crop of *Brassica oleracea* may remove 150 kg Ca per ha. A continuous cropping may result in the reduction of exchangeable Ca in soil. Banana and pineapple crops with yield levels of 40 to 50 t/ha may remove 120 to 140 kg of Ca and Mg . As a thumb rule, S removal per tonne grain production can be taken as 3-4 kg for cereals, about 8 kg for pulses, about 12 kg for oil seeds and 18 kg for cruciferous and 38 kg for mustards. In most of the crop species, the critical limits of S in plants are 0.20 to 0.25%. Plants use approximately as much S as P.

Consumption of micro nutrients by plants

High crop yields remove substantial amounts of micronutrients from the soil, especially Zinc and Boron. Micronutrients depletion in soil depends on soil fertility level and crop yields. Maize based cropping sequence depletes the maximum micronutrients form soil, especially Zn and Fe. The deficiencies of Zn and B are prevalent in most soils especially red and laterite soils.

Nutrient interactions in plants and soils

The greatest competition that occurs between ions with similar size, valency and ion charge. Calcium, magnesium and potassium ions are quite similar in size and charge and hence, exchange sites cannot distinguish the difference between the ions. Often, they indiscriminately accept either ion regardless of which ion is meant for that site. Generally, the binding strengths of potassium and calcium are much stronger than magnesium and they easily out-compete magnesium at exchange sites.

Interaction can be defined as the influence of an element upon another in relation to growth and crop yield. There may be positive or negative interaction of nutrients occurs either in soil or plant. The positive interaction of nutrients gives higher crop yield and such interactions should be exploited in increasing the crop production. Conversely, all negative interactions will lead to decline in crop yield and should be avoided in formulating agronomic packages for a crop.

The knowledge about interactions occurring in soils or plants or both is basic to help develop appropriate and efficient technologies. Further this will help to refine the existing ones to increase agricultural production.

There are mainly two types of interactions effect, *viz*., antagonistic and synergistic effects. Antagonistic effect means an increase in concentration of any nutrient element will decrease the activity of another nutrient (negative effect). While synergistic effects means an increase of concentration of any one nutrient element will influence the activity of another nutrient element (Positive effect). Examples of synergism

Optimum supply of nitrogen ensures optimum uptake of potassium as well as phosphorus, magnesium, iron, manganese and zinc from the soils. The enzyme nitrate reductase which catalyzes the formation of proteins is activated by potassium.

- Optimal levels of copper and boron improve nitrogen uptake by plant.
- Optimal levels of molybdenum improve utilization of nitrogen as well as increases uptake of phosphorus.
- Optimal levels of calcium and zinc improve uptake of phosphorus and potassium.
- Optimal levels of sulphur increases the uptake of manganese and zinc.
- Optimal levels of manganese increases uptake of copper.

One must understand how the negative or positive interaction takes place within or outside the plant. The following nutrient ratios in plants are of utmost importance!

N/P, N/K and N/S: Imbalances of these ratios may depress not only yield levels but also their quality.

The following antagonistic effects have been well established on the uptake of micronutrients by crops:

1. Excess of P adversely affects utilization of Zn, Fe and Cu
2. Excess of Fe adversely affects utilization of Zn and Mn
3. Excess of Zn, Mn, and Cu induces Fe-deficiency in crops
4. Excess of S and Cu induces Mo-deficiency in crops
5. Excess of Lime induces deficiency of all micronutrients.
6. Presence of carbonate and bicarbonate ions in soil due to sodicity or over liming reduces the availability of micronutrient cations to crops which suffer most iron deficiency.
7. Lime X P, Lime X Mo, Mo X P, and Na X K are common negative interactions.
8. Excess of Ca may induce P deficiency
9. Excess of Ca and Mg may depress K uptake
10. Excess of Ca may reduce Mg uptake, if ratio is wider than 7:1
11. Excess of K and NH+ may reduce Mg uptake
12. Excess of N, K and Ca may reduce B toxicity
13. Excess of N,P,K may induces Cu deficiency
14. Excess of NO_3-N may cause Fe deficiency
15. Excess of P and K often produce "bitter" fruit/end product.

Chapter 4

Transformations and Availability of Nutrients in Soils

N-Transformation and availability in soils

a. N-Cycle

b. N transformation in soils

Phosphorus transformation & availability in soils:

a. P-Cycle

b. P transformation in soils

K-transformations& availability in soils:

a. K-Cycle

b. K-Transformations

Calcium, Magnesium & Sulphur availability in soils:

a. Ca&Mg Cycle in soils

b. Ca& Mg Transformations

Sulphur availability & transformations in soil

a. S-Cycle

b. S-Transformations

Micronutrients availability and their Transformations in soil

N-Transformation in soils

N-Availability in soil

Dynamics and transformation of nitrogen in soil is very important with respect to plant nutrition. A bulk of total N is present in the organic form (98%) and only about 2% in inorganic form. However there are continuous transformations between these two pools. The crops utilize nitrogen in the inorganic forms only such as NO_3-N and NH_4-N. The inorganic form of N is also liable to undergo different types of loses like runoff, ammonia volatilization, leaching, denitrification and fixation by clay minerals.

Nitrogen cycle in soil

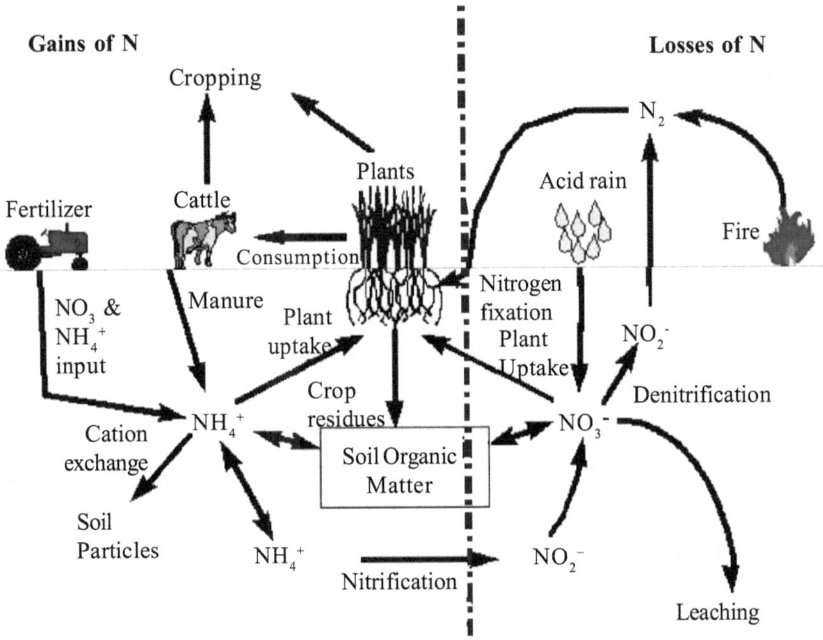

Nitrogen cycle

Nitrogen Cycle

The cycling of N in the soil – plant – atmosphere system involves many transformations of N between inorganic and organic forms.

N transformations in aerobic soils

In aerobic soils, NO_3-N is the dominant form of available N. The mineralization/ transformation of added or native organic matter in soil proceeds up to nitrification stage and giving predominantly NO_3-N and small amount of NH_4-N. There is a quick transformation of NH_4^+ to NO_3-N in the aerobic soils occur which will be utilized by plants. Any fertilizer containing ammonium nitrogen when added to soil gets dissociated to NH_4^+ which readily gets oxidized to NO_3^- ion which is either taken up by the crop or leaches down to the lower horizon as it is readily soluble in water. Some amount of NO_3^-N is also immobilized by soil microbes during the process of mineralization of organic matter.

The organic form of N, particularly hydrolyzable form is slowly mineralized and is transformed to mineral nitrogen through following process

a. Aminization
b. Ammonification
c. Nitrification
d. Denitrification

Nitrogen mineralization

Mineralization is the conversation of organic N to inorganic forms of N as a result of microbial decomposition. Mineralization increases with a rise in temperature and is enhanced by adequate, soil moisture and a good supply of O_2. Mineralization of organic N involves in two reactions.

Aminization

Aminization is the decomposition of proteins and the release of amines, amino acids and urea. A large number of soil microorganisms bring about this change.

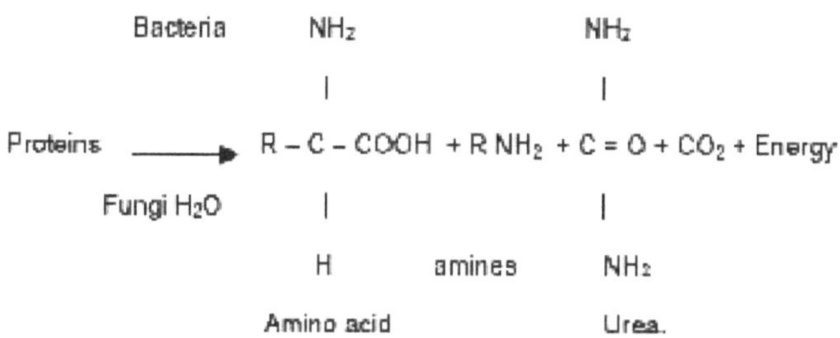

Under aerobic condition the major end products are $CO_{2,}(NH_4)_2$ SO_4 and H_2O.

Under anaerobic conditions the end products are ammonia, amides, CO_2 and H_2S.

The organic compounds and proteins are mainly decomposed by various species of microbes.

Conversion of urea

Urea is a product of aminization. The hydrolysis of urea by the action of urease enzyme is effected by *Bacilli micrococcus, Pseudomonas, Clostridium, Acromobactor* and *Corynebactor*.

$CO\ (NH_2)_2 + H^+ + 2H_2O\ 2NH_4^+ + HCO_3^-$

$NH_4^+ NH_3 + H^+$

$2NH_3 + H_2CO_3\ (NH_4)_2\ CO_3NH_4^+ + CO_3^-$.

The optimum water holding capacity for these reactions is 50-75% and optimum temperature is 30-50°C.

Ammonification

"Amines and Amino acids produced during aminization of organic N are decomposed by other heterotrophs with release of NH_4^+ is termed Ammonifcation

$$R - NH_2 + H_2O \rightarrow NH_3 + R - OH + \text{energy}$$

$$\downarrow H_2O$$

$$NH_4^+ + OH^-$$

Nitrification

Nitrification is the process of biological oxidation by which the ammonical (NH_4^+) form of N converts to nitrate (NO_3^-) form of N. There are two steps.

(a) NH_4 is converted first to NO_2^- and then to NO_3^-.

Nitrosomonas

$$2NH_4 + 3O_2 \rightarrow 2\,NO_2^- + 2\,H_2O + 4\,H^+$$

(b) In the second reaction NO_2^- is further oxidized to NO_3^- by Nitrobactor

$$2\,NO_2^- + O_2 \xrightarrow{\text{Nitrobactor}} 2\,NO_3^-$$

Nitrosomonas are obligate autotropic bacteria that obtain their energy from the oxidations of N and their C from CO_2.

Denitrification

Denitrification is the biochemical reduction of NO_3-N or NO_2-N to gaseous N forms, either as molecular Nitrogen or an oxide of Nitrogen. The most probable bio chemical pathway is

NO_3	$\rightarrow$	NO_2	$\rightarrow$	NO	$\rightarrow$	N_2O	$\rightarrow$	N_2
Nitrate		Nitrite Nitrogen		Nitrous monoxide		Nitrous oxide		Dinitrogen

This is loss mechanism of nitrogen happening in anaerobic soil conditions.

Nitrogen immobilization

"Immobilisation is the process of conversion of inorganic N (NH_4^+ or NO_3^-) to organic N and it is basically the reverse of N mineralization". By this process plant available N forms are converted to unavailable organic forms.

The Microorganisms accumulate NH_4- N and NO_3- N in the form of protein, nucleic acid and other complexes. If C:N ratio is wider than 30, it favors immobilization and lesser C:N ratio encourage mineralization.

Losses of nitrogen

The major losses of N from the soil are due to (1) crop removal and leaching, however under certain conditions inorganic N ions can be converted to gases and lost to the atmosphere. The primary pathway of gaseous N losses are

1. Denitrification
2. NH_3 volatilization.

Nitrogen fixation

The conversation of atmospheric nitrogen to plant available forms readily usable by biological process mediated by microorganisms.

Nitrogen transformation in anaerobic soils

The N-transformations from added or native sources stops at NH_4^+ stage, since nitrification is not possible due to lack of oxygen. When NH_4^+-N containing fertilizers are added to soil, then NH_4^+ is oxidized to NO_3^- in the top layer (aerobic) of flood water or oxidized layer and later which moves down to reduced layer. On the reduced layer NH_4^+-N remains as NH_4-N only for plant uptake. If NO_3-N exists, then it moves down to the reduced layer, where it undergoes denitrification by bacteria (NO_3-NO_2-N_2). The denitrified NO_3^- is lost from the soil to the atmosphere in the form of N_2O or N_2.

If NH_4^+ ions present in the flood water it is subjected to volatilization to atmosphere as NH_3 (ammonia gas) because of higher partial presence of CO_2 and high pH value developed due to alkalinity. Therefore, there is an accumulation of NH_4^+ ions in the reduced layer, which is either absorbed by root or gets oxidized in the rhizosphere to NO_3^- ions and are lost due to dentrification. Some of the NH_4^+ and NO_3^- ions also get immobilized by the soil microbes & some NH_4^+ may get fixed by the clay lattice.

Estimated quantities of Gain sources (A)

A_1 Symbiotic N fixation by rhizobium bacteria: 30-250 kg ha^{-1}

A_2 Nonsymbiotic N fixation by azotobacter SP: 3-15 kg ha^{-1}

A_3 Mineralization of organic forms: 2 t/ha annually

A_4 Rainfall (precipitation): 5-20 kg ha^{-1}

Estimated quantities of losses (B)

B_1 Crop removal: appriximately 20-60 % of the applied N is taken up by yields (15-100 kg ha^{-1}, 200 kg for totalbiomass)

B_2 Ammonium fixation: 3-18% of the total N

B_3 Ammonium volatilization: 5-15% of the applied N B4 Denitrification: 5-15 kg ha^{1}

B_5 NO_3^- Leaching 5-10 % of the applied N

B_6 Erosion depending on rainfall and its intensity: 5-50 kg ha^{-1}

Amounts of losses are influenced by several factors; among these soil characteristics, climatic conditions as well as cropping systems and agrotechnics play important role in the extent of losses.

Effects of ammonium-N and nitrate-N supply on plants

Due to their paramount significance in plants' lives the supply of ammonium-N and nitrate-N has various important effects.

1. The reduction of NO_3^- is an energy-requiring process (reduction of each NO_3^- ion requires 2 molecules of NO_3^- reductase for proteinsynthesis).
2. When plants take up high levels of NO_3^- N, an increase in cation (K+, Ca++, Mg++) absorption willoccur.
3. High levels of NH_4^+ -N may be toxic for cells and retard growth. Due to the ion antagonism, it restricts K uptake.
4. Cereals, corn, rice, pineapple use both forms of N while potato, tomato and other solanaceae crops prefer a high nitrate/ammonium ratio for optimum growth.

Phosphorus transformation & availability in soils

Phosphorus cycle in soil

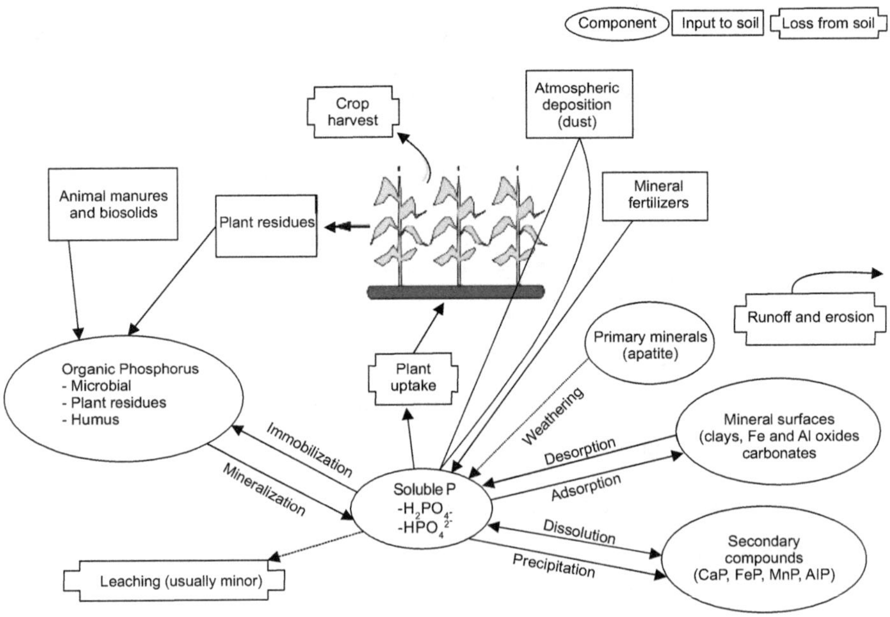

In P cycle the phosphorus is not involved in any exchange process with the atmosphere. The amount of dissolved or solution P mostly $H_2PO_4^-$ and HPO_4^{2-} ions in the soil is very small and crops will utilize P from this source. The soil solution P can come from mineralization of organic matter, added fertilizers, adsorbed phosphate ions and solid P compounds (primary & secondary minerals). Any P present in solution P form and it is available to plants is called Labile P, where as any P which is bounded in solid P compounds such as primary and secondary minerals is called Non Labile P which is not available to plants.

P transformation in aerobic soils

P transformation depends on their solubility, interaction with other soil components. When a water soluble phosphate fertilizer, such as super phosphate is applied to a soil, P dissolves immediately and enters the soil solution and readily forms new compounds with Ca as Ca-PO_4 and carbonates of hydroxyapatite in calcareous and alkaline soils respectively. While in acid soils rich in Al, Mn and Fe, applied P is converted as Al-PO_4, Mn – PO_4 and Fe-PO_4. All these gets newly formed compounds are precipitated thus reduce the availability of P to crops.

P transformation in anaerobic soils

P- Transformation in submerged or flooded conditions/soils with a wide range of pH create reducing conditions which lead to reduction of Ferric-PO_4 to ferrous PO_4 resulting in greater availability of P in the reduced soil. Organic acids formed under submerged conditions also solubilize PO_4.

Estimated quantities of sources (A)

A_1 Soil solution P: 0.03 – 0.2 mg liter^{-1} (200- 6000 kg ha^{-1} in the 0-25 cm layer)

A_2 Eulibrium in organic/inorganic P forms: between C/P ratio: 200-300

A_3 Farmyard manure: 60-120 kg/ha P_2O_5

Estimated quantities of losses (B)

B_1 Crop removal: Approximately 25-70 kg ha^{-1} P2O5 (10-30 kg ha^{-1} P) is taken up by yields – approx. 80 percent of outputs.

B_2 Runoff: approximately 0.25 kg ha^{-1} P annually.

B_3 Erosion: 0.2-0.8 kg ha^{-1} P with 1 ton of soil loss B6 Erosion is depending on rainfall and its intensity.

Important

- Amounts and ratios of $H_2PO_4^-$ and HPO_4^- depends on soil pH & Equilibrium. Maximum P availability exists at pH range between 6.5 -7.0.
- Inorganic P in solution can be adsorbed to mineral surfaces or precipitated (often referred as P fixation or retention).
- Residual P availability can persist for several years (depending on P rate, crop removal, soil properties *etc*).
- Mineral fertilizers are increasingly important as sources of soluble phosphorus.
- There are no significant losses of phosphorus caused either by leaching or in other forms except sandy soils with extremely light texture and poor in SOM.
- Some 1-2% is in microbial tissue, and only 0.01% exists as soluble phosphorus.

K-transformations & availability in soils

K- Availability

Potassum is present in soil solution as K^+ion which is readily available to plants. But this form is in dynamic equilibrium with exchangeable K which intern with fixed K. Fixed K is in equilibrium with mineral K.

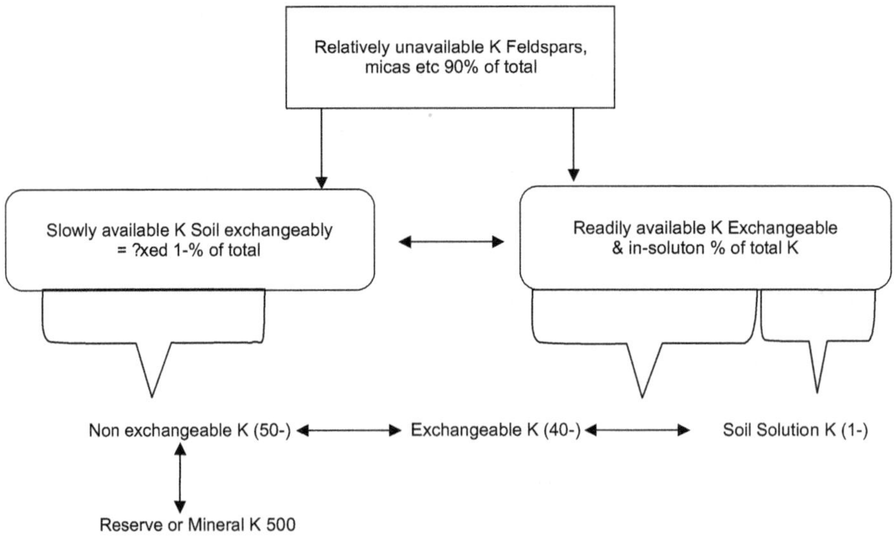

Fig. 4.1: Forms & Availability of K

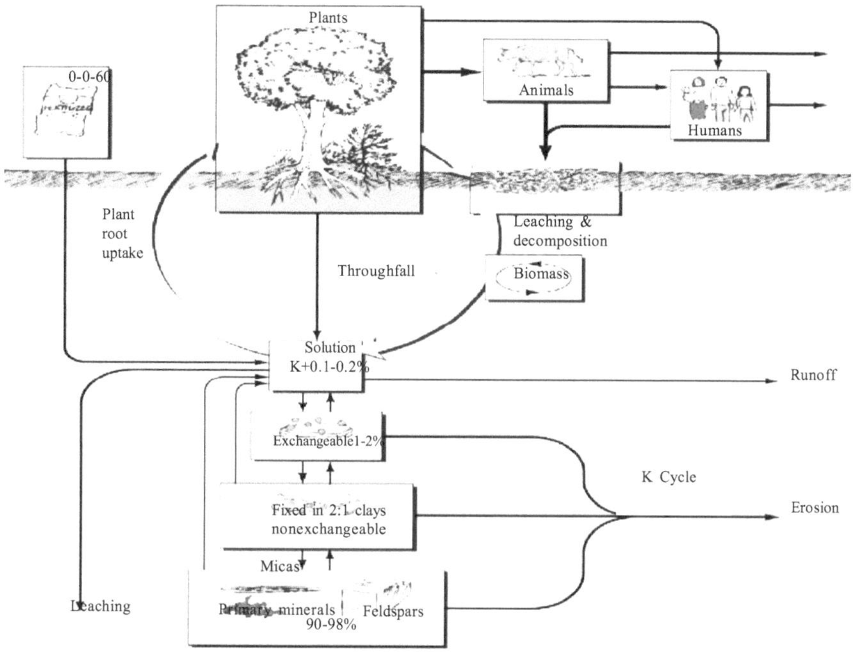

Fig. 4.2: *K Cycle*

K- Transformations

When potassic fertilizers are added to the soil, K may either remain in soil solution or in exchangeable form on the clay surface or in non-exchangeable form held by illitic clay minerals as fixed K as it is not available to plants. Plants get K mainly from solution-K and exchangeable K. As and when the exchangeable K fraction is depleted substantially or exhausted by cropping, the non exchangeable form of K replenishes the exchangeable K and supply of K to plants is maintained and equilibrium is attained. The reservoir source of K is mineral lattice K which undergoes weathering and releases K. In all these transformations equilibrium is maintained always.

Estimated quantities of sources (A)

A_1 Soil solution K: 0.2 – 10.0 mg liter^{-1} K.

A_2 Release from weathering of clay minerals (layer silicates such as feldspars, micas *etc*).

A_3 Atmospheric deposition: 2-6 kg ha^{-1} K_2O.

A_4 Mineral fertilizers - Depending on crop yield requirement.

A_5 Farmyard manure: 120-200 kg ha^{-1} K_2O (0.2-2.0 % K in DM).

Estimated quantities of losses (B)

B_1 Crop removal: approyimately kg ha^{-1} K_2O taken up by yields Taken up by yields – approx. 70 per cent of outputs

B_2 Fixation by clay minerals (vermiculites, smectites *etc.*) may reach 1000 kg ha-1 K_2O

B_3 Leaching: approximately 1.6 kg ha^{-1} K annually, depending on soil texture

B_4 Erosion is depending on rainfall and its intensity

Important

- Primary and secondary minerals are original sources of K
- Organic matter holds and releases some of the K
- Interaction with soil colloids exchange able and nonexchange able (fixed) forms of potassium is in equilibrium
- The bulk of the potassium is held in the primary and secondary (clay)minerals.
- Fixation by layer silicates strongly increases with drying of the soil: dry fixation may be 5 times higher than in moistsoils

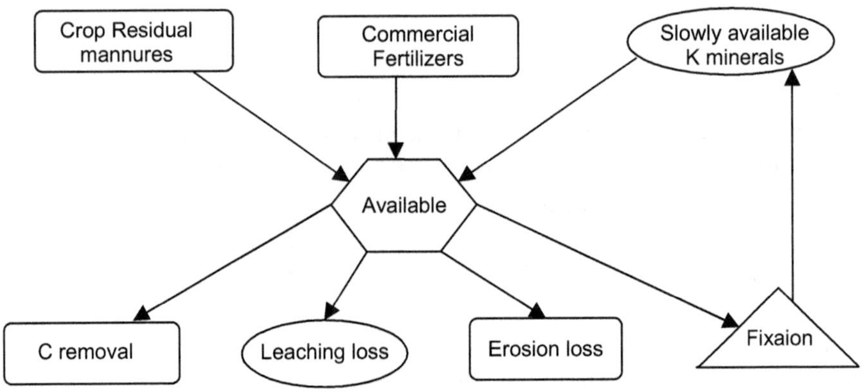

Fig. 4.3: Gains & Losses of Available K in Soil

Calcium and magnesium availability in soils

Ca & Mg availability in soils

Ca and Mg are the most abundant cations occupying the exchange sites of the soil colloids of both inorganic (clay) and organic (humus). Thus most soils contain enough Ca and Mg except highly weathered leached acid soils and alkali soils. Deficiencies of Ca and Mg frequently occur in coarse textured as well as acidic soils developed under high rainfall conditions. The critical limits of exchangeable Ca and Mg vary widely among soils. However average value of <2.0 m.eq/100g for exchangeable Ca and < 0.5 m.eq/ 100g for exchangeable Mg are considered critical limits for availability.

Ca & Mg transformations

Ca & Mg occupying the exchange sites of the soil colloids (clay & humus) are subjected to cation exchange reactions then released into soil solution for plants absorption or adsorbed on the clay and organic matter surfaces. Soils usually contain less Mg than Cabecause Mg^{2+} ions are not adsorbed as strongly by clay and organic matter as Ca^{2+} ions and further Mg^{2+} ions are more susceptible to leaching than Ca^{2+}ions.

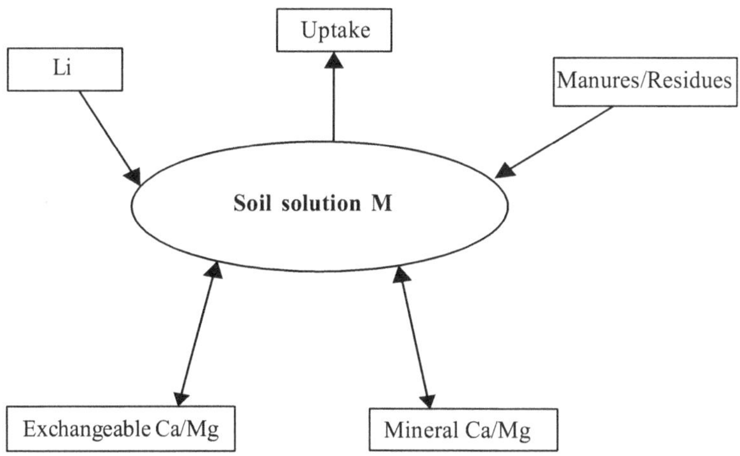

Simple Representation of Ca & Mg Equilibrium & Transformations in Soil

Sulphur cycle and transformation in soils

Sulphur cycle in soil

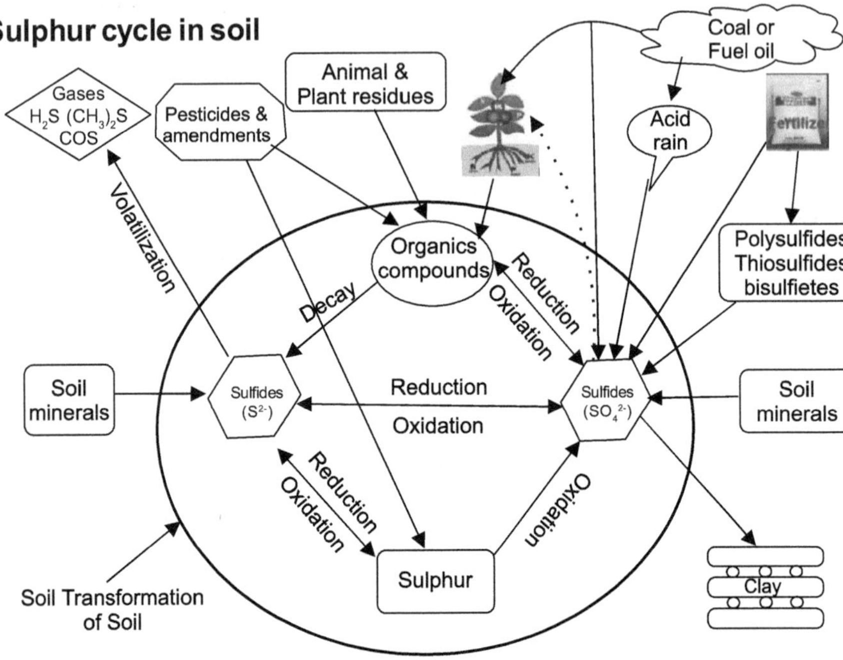

Sulphur cycle

S-Transformations

A) In aerobic soil

The sulphur in soil is being cycled continuously between inorganic and organic forms just like Nitrogen. The nature of compounds formed and their transformations are influenced by the biologically mediated process. Organic matter is the major of source of S, it undergoes oxidation to SO_4^{2-}. This is brought about by autotrophic bacteria belonging to the genus *Thiobacillus* under aerobic conditions of soil.

$$\text{Organic S} \xrightarrow{\text{Aeration } (O_2)} SO_4^{2-} + H^+$$

(Proteins, Amino acids). *Thiobacillusthioxidans*

$$\text{Fe S} \quad SO_4^{2-} + H^+ \xrightarrow{\text{Aeration } (O_2)} \text{Pyrite}$$

Thiobacillusferroxidans

The released SO_4^{2-} ion is taken up by plants and microorganisms. The rate of mineralization is affected by factors such as moisture (60% field capacity), aeration, temperature (35-40°C) and Soil pH.

B) **In anaerobic soil**

Under anaerobic conditions or flooded soils, the soil aeration is completely cut off. The SO_4^{2-} ion is reduced to sulphides (S^{2-}) by the autotrophic bacteria such as *Desulpho vibrio and Desulphomaculans* which use SO_4^{2-} for their O_2 requirements.

$SO_4^{2-} + 8 H^+ + 8^{e-} = S^{2-} + 4H_2O$.

Micronutrients cycle and their transformations in soil

The total content of Fe, Mn, Zn, Cu, Co, Cl& B varies considerably in different soils. Except Zn, Cu and B, all other micronutrients (Fe, Mn, Cl, Mo) are present in Indian soils in sufficient amounts to sustain agricultural productivity. Zn and Boron deficiency is found in all the soils of agro ecological regions of the country.

The availability of micronutrient cations in soil is highly affected by inorganic ions in soil solution, soil solid constituents like free oxides of Fe & Al, soil organic matter, fertilizers and amendments applied to soil

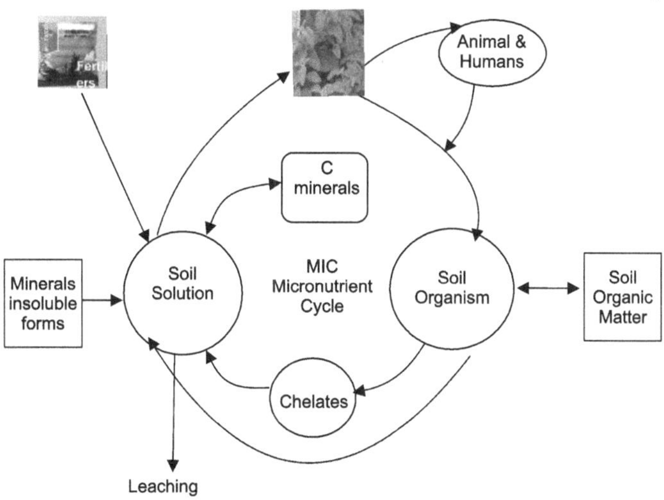

Fig. 4.4: Micronutrients Cycle

Table: Soil conditions resulting shortages of available nutrients and Inducing nutrient deficiencies for crops

Nutrient	Soil Conditions resulting Shortages in availability - inducing deficiency
N	Excess leaching with heavy rainfall low organic matter content of soils, burning the crop residue.
P	Acidic, organic, leached, and calcareous soils, high rate of liming.
K	Sandy, organic, leached, and eroded soils, high liming application, intensive cropping system.

Ca	Acidic, alkali, or sodic soils.
Mg	Similar to calcium.
S	Low* organic matter content of soils, use of N and P fertilizers containing no sulfur, burning the crop residue.
Fe	Calcareous soils, soils high in P, Mn, Cu or Zn, high rate of liming.
Zn	Highly leached acidic soils, calcareous soils, high levels of Ca, mg, and P in the soils.
Mn	Calcareous silt and day, high organic matter, calcareous soils.
B	Sandy soils, naturally addic leached soils, alkaline soils with free lime.
Mo	Highly podzolized soils, w'ell-drained calcareous soils.

Source: Fageria *et al*., 1991.

Interrelations in the soil-crop-nutrient system

The most important relationships are;

Firstly, there are soil factors. Ion exchange is a crucial process in the nutrient supply of plants. It is a reversible process in which ions are exchanged with other ions (having the same charge) from the soil solution. Ion exchange can take two forms cation exchange and anion exchange. Cation Exchange Capacity and Anion Exchange Capacity are among the most important soil chemical properties influencing nutrient availability and retention in soils (expressed as milliequivalents of (-) charge per 100 g of dried soil = meq 100 g-1). To refer to features common abbreviations are used: CEC means Cation Exchange Capacity while AEC represents anion exchange capacity.

It is very Important to note here that the properties of the ions determine the strength of adsorption and conditions of desorption.

Another of soil factors, which is also one of the most important relationships is Buffering Capacity, often referred to by the abbreviation BC. As plant roots take up ions from the soil, adsorbed and exchangeable ions will be desorbed from the exchange sites. The concentration in the soil solution is maintained by the buffering capacity of the soil. Relatively constant ion concentration in the soil solution can be maintained by resupplying ions to the soil solution. The buffer capacity of various soils depends on the cation exchange capacity (CEC) and soil organic matter (SOM). Consequently, the quantity of clay minerals and humus will determine the extent of buffering. Soils containing dominantly 2 :1 clay minerals have higher buffer capacity (BC), while sandy soils are poorly buffered.

Table: Ion Adsorption Capacity of Several Surfaces CEC and AEC meq per 100 g dry weight

Soil particle	CEC meq 100 g-1	
Montmorillonite	80-120	
Caolinite	3-15	
Illite	20-50	
Vermiculite	100-150	
Humus/ SOM	150-250	
Plant roots	**CEC AEC meq 100 g’^{1}**	
Cereals	10-25	60-80
Com	13-29	140-160
Potato	3540	
Tomato	60-62	
Pea	50-60	150-160
Weeds	± 100	

Important

- Generally it was observed that dicotyledons have higher CEC than monocots.

Table: Typical CEC Values of Different Soil Textures CEC meq per 100 g dry weight

Soil Texture	CEC meq 100 g^{-1}
Sands (Tight colour, extremely poor OM)	3-5
Sands (dark colour, poor OM)	10-20
Loams	10-15
Silt loams	15-25
Clay loams and clay	20-50
Organic soils	50-100

Havlim *et al.*, 2005

The concept of the supplying capacity of soil nutrients to crops

The quantities of soil nutrients available to plant roots always depend largely on several soil and other factors such as climatic and other environmental conditions *etc*. Nevertheless, these quantities are also influenced by the intensity level of fertilization (*i.e.* amounts added to soils) and nutrient transformation processes. Soils have their characteristics in nutrient dynamics: retention, fixation and their capacities in supplying available nutrient forms to root uptake.

It is important to understand that nutrient transformations in soils exist in dynamic equilibrium between mobilization and immobilization. These processes include both mineralization or decomposition of organic substances called as soil organic matter (SOM) and chemical fixation as well as biological fixation by soil microorganisms.

Mobilization	$\rightleftarrows$	Immobilization
Mineralization.		Precipitation (insoluble compounds)
Decomposition		Biological fixation

Amounts of nutrients to be mobilized (A)

Readily available – nutrient potential (B)

Nutrient capacity (A+B) (C)

Nutrient intensity – rate of transformation (w)

The different letters denote the following:

The amounts of nutrients to be mobilized are represented by (A),

The readily available – nutrient potential is referred to as (B),

The nutrient capacity, which is (A+B) is shown by (C).

Nutrient intensity – rate of transformation is represented by (w).

These processes can be interpreted by the following, simplified equilibrium:

w1 = rate of mobilization

w2 = rate of immobilization

Important

- Soil nutrient transformation processes are *reversible.*
- Equilibrium exist when rate of mobilization and that of immobilization is the same, i.e. W1 = W2.

Chapter 5

Role of Microorganisms in Organic Matter Decomposition in Soil and Importance of pH in Plant Nutrition

Organic matter and its composition

Decomposition

Types microorganisms involved

Enzymes involved in decomposition process?

Major Process of organic matter decomposition

1. Mineralization

A. Nitrogen mineralization

a. Aminization
b. Ammonification
c. Nitrification
d. Denitrification

B. Phosphorus mineralization

C. Sulphur mineralization

2. Immobilization

Rate of decomposition of organic compounds

Products of Decomposition of organic matter

Factors affecting decomposition

Formation of humus

Fractions and properties of humus

C: N ratio and its importance

Importance of pH in plant nutrition

Organic matter

Soil organic matter is a complex matter obtained from the plants and animal residues containing both organic and inorganic chemical compounds.

Composition of organic matter

A. Organic compounds

1. Nitrogenous compounds eg: Amine, amide compounds and amino acids (water soluble) proteins (simple, combined, derived), peptone, purine, peptide, nucleoprotein, protease, pyridine, alkaloid, pirimidin.
2. Non nitrogenous compounds

   Eg:- Carbohydrates (mono, di, poly and oligosaccharides, amino sugars, sugar acids, sugar alcohol), lignin (aliphatic hydroxyl and carbonyl group found), Cellulose and hemicelluloses, tannin, fat, wax and resin, pigments, organic acids (acetic, oxalic, saccharic, propionic, benzoic, pomitic), organic phosphorus compounds (nucleic acids, phospholipids, inositol phosphate) and organic sulphur compounds (cystein, cystine, methionine).

B. Inorganic compounds

Organic matter contains several inorganic elements such as H, O, N, P, Ca, Mg, S, Na, K, Fe, Al, Mn, Zn, Mo, Si, B, Co, Cl etc. All these are in metallic complex form hence organic matter is water insoluble and is not destroyed by leaching.

Decomposition

It is the decaying or rotting of organic materials by various groups of microorganisms and enzymes and converted into simple inorganic elements or compounds. It improves softness of the materials for further mineralization process. Chemically it is a burning or oxidation process.

Two types of decomposition are:

1. Aerobic decomposition- oxygen is required
2. Anaerobic decomposition- oxygen is not required.

Types of microorganisms involved in organic matter decomposition

Decomposition is a purely microbial process involves several species of microorganisms. They are; Bacteria, actinomycetes, fungi, protozoa and supporting organisms like insects, worms, rodents, termite, ants etc.

All the soil organic matter is not decomposed by the same group of microorganisms. At different stages of decomposition different species of microorganisms enter into the degradation process. As the degradation proceeds newer materials are synthesized by soil microorganisms.

Enzymes involved in decomposition process

Cellulase-breaks cellulose

Urease – breaks urea $CO(NH_2)_2$ to CO_2 and NH_4

Phosphatase – breaks humus $–O-P-(OH)_2$ bond to produce humus, OH & H_3PO_4

Sulfatase-breaks the humus O-S-OH bond to produce humus, OH & H_2SO_4

Protease – breaks bond linking two amino acids to separate amino acids

Decomposition processes

Two major decomposition processes involved

1. Mineralization

2. Immobilization

1. Mineralization: is the conversion of an element from organic form to an inorganic/mineral form is called mineralization. Mineralization occurs for each element present in the organic matter individually especially for N, P and S

A. Nitrogen mineralization

It is the conversion of organic nitrogen into inorganic Nitrogen NH_4 and NO_3 by microorganisms

Organic N $\rightarrow$ ProteosePeptone $\rightarrow$ Amino compounds

(proteins and amino acids) $\downarrow$
ammonia->nitrite nitrate $\rightarrow$

Different steps involved in N-mineralization

1. **Aminisation**: proteins and other complex nitrogenous compounds are converted into amino acids and amino compounds by the action of enzymes and microorganisms, CO_2 and energy is released.

2. **Ammonification**: it is the biochemical conversion of amino compounds and amino acids into ammonia by bacterial decomposition.

Enzyme

Amino acids + Water → Ammonia + energy + amino compounds hydrolysis

Or

$R\text{-}NH_2 + HOHR\text{-}OH + NH_3 + \text{energy}$

This process is governed by aminase and deaminase enzymes. The liberated NH_3 is utilized by plants.

3. **Nitrification:** It is the biochemical oxidation of ammonia into nitrite and then finally into nitrate. This process carried out by by autotropic bacteria (*Nitrosomanas* and *Nitrobacter*) in aerobic condition. Two steps are involved.

Nitrosomanas and N*itrosococcus*

$2NH^+_4 + 3O_2 2NO_2 + 2H_2O + 4H+ +\text{energy}$ Enzymatic oxidation (unstable)

$$2NO_2^- + O_2 \xrightarrow[\text{Enzymatic oxidation (available)}]{\textit{Nitrobacter}} 2NO_3^- + \text{energy}$$

Rate of nitrification depends on suitable temperature, humidity, pH, season, aeration, addition of lime (increases), nature of organic matter etc.

4. **Denitrification:** It is the biochemical reduction of nitrate or nitrite to gaseous nitrogen (N_2), either as molecular N or oxides of N.

Anaerobic decomposition

$NO_2^-/NO_3^- N_2 \quad \rightarrow \quad \uparrow$

Reduction

Loss of nitrate by reduction and assimilation

B. Phosphorus mineralization

The process of conversion of organic forms of P into inorganic forms of P by P decomposing microorganisms especially by mycorrhiza species is known as P-mineralization.

The organic-P found in organic manures mainly as nucleic acids, phytin, phospho lipids, inositol PO_4, lecithin etc. The organic P is not directly available to plants and it has to undergo decomposition by micro organisms especially mycorrhyzal species.

$C_6H_{24}O_{27}P_6 + 6O_2\ 6H_3PO_4 + 3H_2O + 6CO_2$.

(Phytic acid)

C. Sulphur mineralization

The organic forms of S compounds such as methionine, cystine, cysteine are converted into inorganic sulphate forms (available) by aerobic bacteria species especially *Thiobacillus* and *Thiobacter*.

1. Cystine + Water ———————-Pyruvic acid $+H_2S+NH_3$
   Oxidised

2. H_2S ——————————————SO_4 (sulphate)

Immobilization: the conversion of an element from the inorganic to the organic form in microbial tissues or in plant tissues, thus rendering the element not readily available to other organisms or to plants.

Rate of decomposition of organic compounds

Rate of decomposition varies with the types of organic compounds, some may undergo decomposition very fast and others may very slow. The following is the decreasing order of rate decomposition:

Starch , sugars and simple proteins > Crude proteins> Hemi cellulose> Cellulose > Lignin > fat > Fatty compounds and waxes etc.(Most resistant)

When fresh organic manures are added to soil 60% of it decomposed within the first 6 months, 20% in the next 3-4 years and remaining 20% decomposed beyond 4 years. This results in accumulation of lignins, fats and waxes in large quantities.

Products of decomposition of organic matter

Under well drained, aerated (oxidised) soil: CO_2, NO^-_2, NO^-_3, H_2O, PO^-_4, SO_4^{2-}, $H_2$and other essential plant nutrients. Antibiotics, Auxins, hormones, phytohormones.

Under anaerobic condition (water logged and compacted soil): Methane (CH_4) (swamp gas), organic acids (R-COOH, NH_4 and amine groups ($R-NH_2$), Toxic gases like H_2S, dimethyl sulphide and ethylene ($CH_2=CH_2$) *etc.*

Factors affecting decomposition

The most important conditions that affect the rate of decomposition are

1. **Temperature:** Cold periods retard the organic matter decomposition and there will be more accumulation of organic matter on the top soil compared

to that of warm climates. The most suitable temperature is 30-40° C for proper decomposition.

2. **Soil moisture:** Near or slightly wetter than field capacity moisture conditions are most favorable for decomposition. About 60-75% water holding capacity (WHC) is optimum.
3. **Soil pH**: 6-8 pH or neutral pH is required for optimum growth of microorganisms. Bacteria at 6 - 7 pH, Actinomycetes is more at pH 8 -10, Algae pH of 5.5-7.5, Fungi- pH 4.0, Protozoa – pH 3.0
4. **Nutrients:** Lack of nutrients, particularly N reduces microbial growth and it slows decomposition. Addition of nutrients by N fertilizers (urea) increases the speed of decomposition
5. **Soil texture**: Soils higher in clays tend to retain larger amounts of humus, other condition being equal.
6. **Aeration**: Good aeration increases the rate of decomposition and supply oxygen.
7. **Nature of plant matter**: composition and age of plants and vegetations affect much their decomposition. It is fast in young, tender and juicy material, But slow with more cellulose and hemicelluloses content.

Formation of humus

Definition of humus

Humus is defined as they are more or less stable fraction of the soil organic matter remaining after the major portions of added plant and animal residues have decomposed. Usually it is dark or brown in colour. They are high molecular weight compounds, complex, resistant, polymeric compounds. They are amorphous and colloidal organic substances.

Composition of humus

Humus is a heterogeneous mixture of complex organic compounds. It is mainly made up of 58% C, 3-6% of N- and C:N ratio of 10:1 to 12:1.

Humus formation

Humic substances are produced when plant residues and other organic debris are broken down and/or chemically altered by microorganisms and subsequently recombine under the influence of enzymes. Humus formation is a complex two stage process in which organic residues of plant and animal origin undergo profound transformation.

1. The decomposition of the original components of tissues and their conversion by microorganisms in to simpler chemical compounds and partially to products of complete mineralization (CO_2, NO_2, NO_3, NH_3, CH_4, H_2O etc.)

Compound	Broken into
Lignin	polyphenols, phenolic acids
Proteins	polypeptides and amino acids
Carbohydrates	simple sugars

2. The synthesis of organic compounds with the formation of high molecular weight humic substances of specific nature viz.High molecular weights humic acids (HAs) and Fulvic acids (FAs).

Fractions of humus: Humus is mainly composed of two major groups, they are

I. Humic group

II. Non-humic group

I. Humic group: The humic substances make up about 60-80% of the soil organic matter. On the basis of resistance to degradation and of solubility in acids and alkalis, humic substances have been classified into five chemical groupings.

1. **Fulvic acid**: lowest in molecular weight, lightest in colour, soluble in both acid and alkali and most susceptible to microbial attack. Contain uronides, simple carbohydrates and their sugars, phenolic glycosides, tannin and other organics, also rich in N and P.
2. **Humic acid :**Medium in molecular weight and colour, soluble in alkali but insoluble in acid and intermediate in resistance to degradation forms largest bulk, responsible for importing its characteristics properties to humus.
3. **Humin:** Highest in molecular weight, darkest in colour, insoluble in both acid and alkali and most resistant to microbial attack. Polymerized product of a part of the FA and HA fractions.
4. **Apocremic acid**
5. **Hematomelanic acid**.

All the five fractions are amorphous and show no signs of crystallization.

II. Non humic group: It comprises about 20-30% of the organic matter in soils. They are less complex and less resistant to microbial attack than those of humic group. They are comprised of specific organic compounds with definite physical and chemical properties.

E.g.:

a. ***Polysaccharides***- Polymers with sugar like structures. They are effective in enhancing soil aggregates stability.

b. ***Polyuronoids***- They are not found in plants but have been synthesized by the soil microbes and held as part of the organisms body tissues.

c. Organic acids and protein like materials.

Properties of humus

1. It is a light bulky amorphous material of dark brown to black colour. The black colour of surface soil is usually due to the presence of humus.
2. It has a great water absorbing and water holding capacity. 100 part humus- 181 part of water.
3. It possesses the power of adhesion and cohesion (but less than clay) so it acts as a cementing agent in crumb formation. In Sandy soils- adhesive capacity and in clay soils – cohesive capacity increases
4. It has a high ion adsorbing capacity (4-6 times that of clay) and CEC is very high (300-350 m.eq./100gm)
5. It is insoluble in water.
6. It behaves like a weak acid and forms salts with bases.
7. It acts as buffering agent and also as an oxidation reduction buffer.
8. It serves as a source of energy and food for the development of various microorganisms.
9. An important source of nutrients for higher plants.

C: N ratio and its importance

C:N ratio

It is defined as the ratio of the weight of organic carbon to the weight of total nitrogen in a soil or organic matter. It is the relationship between organic matter and nitrogen content of soils or plants.

$$\text{C: N ratio} = \frac{\%\ \text{OC}}{\%\ \text{N}}$$

Importance of C:N ratio

1. C:N ratio mainly controls decomposition rate in soil

- The wide C:N ratio leads to slow decomposition rate, nutrient immobilization may occur, carbon and energy wastage in large quantities. Activity of microorganisms restricted total amount of N is limited
- In narrow C:N ratio, Carbon and energy starvation occur. Plant residues decompose quickly and release nitrates readily. Amount of CO_2 released/ unit of carbon decomposed is less as more of it is metabolized and converted into microbial tissues.

When the residue with high C/N ratio is added to soils, there will be intense competition among the microorganism for available N.

The C/ N ratio in residues helps determine their rate of decay and the rate at which N is made available to plants.

Ex : Speed of decomposition becomes slow with more/wide C/N ratio residue or low N percentages. On the contrary low/narrow C/N ratio or high N percentages speeds the decomposition rate.

2. It is a source of food and energy for plants

Soil organisms require carbon for building essential organic compounds and to obtain energy for life process, but they must also obtain sufficient N to synthesize N containing cellular components, such as amino acids, enzymes and DNA. Microbes need to find about 1 g of N for every 24 g of C in their food. Microbes have 8:1 ratio means – microbes must incorporate into their cells about 8 parts of carbon for every 1 part of N.

3. Influence of C/N ratio on N release

It controls N availability in soils/plants: It controls N availability in soils/plants. If C/N ratio of OM is about 25:1, the soil microbes will have to scavenge the soil solution to obtain enough N. Thus, the incorporation of high C/N residues will deplete the soil native N, causing higher plants to suffer from N deficiency. While low C/N ratio (<20) Organic matter helps in increase in N content of soil for plants and organisms.

5. The decay of organic matter can be delayed

If sufficient nitrogen to support microbial growth is neither present in the material nor available in the soil

6. Influence of C/N ratio on soil ecology

The soil ecosystem consists of saprophytic bacteria and fungi and nematodes, protozoa and earthworms that grow rapidly on organic residues as food source.

It is related to release of available N, total organic content and accumulation of humus.

C:N ratio's of some of the organic materials

1.	Alfalfa	20:1
2.	Microbial population	10:1
3.	Soil organic matter	10-12:1
4.	Maize stalk	40:1
5.	Rice straw	100:1
6.	Rye straw	200:1
7.	Saw dust	400:1
8.	Clovers (mature)	20:1
9.	Soil humus	11:1

Ratio varies from about 10 for leguminous and young plant materials and >100 for cereal straws.

Importance of pH in plant nutrition

The soil pH or soil reaction is the chemical properties/characteristics of the soil showing the degree of acidity, alkalinity or neutral condition of soil. The pH is having many roles in crop production and particularly in plant nutrition.

1. It significantly influences other chemical as well as biological properties and also affect the availability of most of the chemical elements of importance to plants and microbes.
2. The soil pH greatly affects the solubility of minerals.
3. The soil pH determines the amount and type of nutrient element availability in soils.

   Eg: In strongly acidic soils (pH 4-5) usually have high and toxic concentration of soluble Al and Mn.
4. The soil pH also influences plant growth by the effect of pH on activity of beneficial microorganisms.

5. Soil pH affects the mobility of many pollutants in soil by influencing the rate of their biochemical breakdown, their solubility and adsorption to colloids.
6. Better nutrient availability found at neutral pH (6.5-7.5) like N, P, K, S, Ca, Mg but in low pH acid soil toxicity of Fe, Al, Mn *etc*. and deficiency of P, Mo *etc*. while in saline and alkaline soils Fe, Mn, Zn and Cu may be deficient. Mo availability is more.
7. It helps in recommendation of soil amendments and fertilizer applications.
8. It is a good guide for predicting which plant nutrients are likely to be deficient.
9. It determines the microbial activity and the rate of decomposition.
10. Availability and mobility of both macro and micro nutrients in soil is greatly affected by soil pH.
11. The pH level affects soil physical properties ex: dispersion of clays and formation of stable aggregate structure.
12. Suitability of soil for crop production.

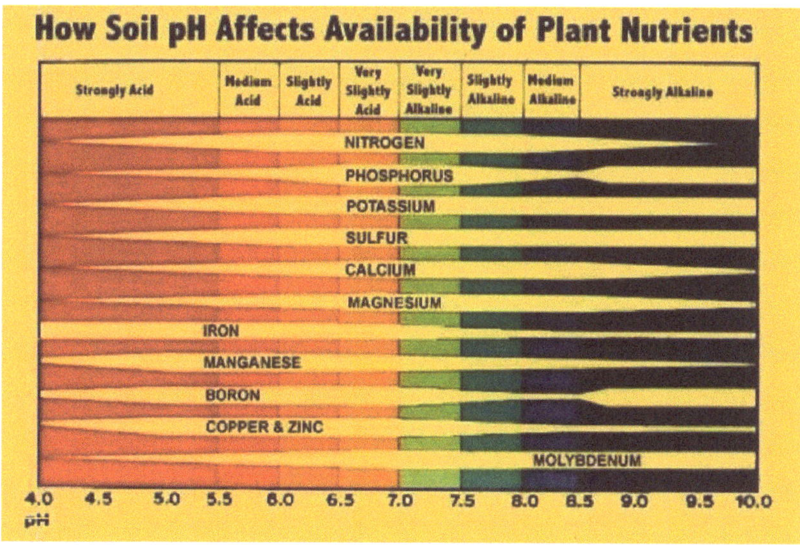

Chapter 6

Acid Soils and Calcareous Soils-Their Formation and Management

1. Acid soil
2. Causes of soil acidity
3. Problem of soil acidity
4. Management or Amelioration of soil acidity
    a. Lime requirement (LR):
    b. Liming reaction in soils
    c. Benefits of liming on acidic soils
5. Calcareous soil
6. Formation, characteristics & management of calcareous soil

Soil acidity is an important agricultural problem while evaluating the production potential of most of the crops. Millions of hectares of land lie idle because of strong soil acidity. In India 49 million hectares of land have problems of soil acidity. Soil acidity is more common in regions with high rainfall. This leads to leaching of appreciable amounts of calcium ,magnesium, potassium and sodium from the surface layer leading to build up of H and Al species causing soil more acidity.

Acid soil

Acid soil is a base unsaturated soil with enough amount of adsorbed exchangeable H^+& Al^{3+}ions with the soil pH of <6.5. An acid soil is actually a mixed H-Al system, *i.e.* such a soil has both H^+ and Al^{3+} ions as exchangeable ions.

Causes of soil acidity

1. Leaching of bases *i.e.* Ca, Mg,K and Na, due to heavy rainfall,
2. Acidic parent material like granite
3. Use of acid forming fertilizers like Urea and ammonium sulphate
4. Decomposition of organic matter leading to release of various organic acids
5. Continuous cultivation of crops resulting in more absorption of basic cations form the soil
6. Acid rains containing sulphate and nitrate ions
7. Aluminium and Iron polymers- One Al species release 3 H^+ ions upon step wise hydrolysis and similar reactions are possible with Fe^{3+}.
8. Laterization, podzolization and accumulation of undecomposed organic matter under marshy conditions contribute to soil acidity.

Problem of soil acidity

Soil acidity causes physical, chemical and biological problems in soil.

1. **Physical problems**- in extreme acid soil, soil will be heavily aggregated due to sequioxides and very compact like laterite.
2. **Chemical problems** are
   a. Acid toxicity
   b. Toxicity of different nutrient elements- such as Fe, Al and Mn which are in more soluble form
   c. Nutrient availability is reduced especially N, P, K Ca, Mg, Na,
   d. Nutrient imbalances due to fixation of PO_4 by Fe, Al and Mn,
   e. Boron and Mo availability decreased
3. **Biological problems**: Microbial activity decreases
   a. Bacteria and actinomycetesare inactivity when the soil pH drops below 5.5.
   b. Nitrogen fixation in acid soils is greatly affected by lowering the activity of *Azatobacter* sp., and *Rhizobium* species.
   c. Fungi can grow well under very acid soils and cause various diseases like Foot rot of Black pepper and blights of potato *etc.*

Management or amelioration of soil acidity

In general the fertility status of acid soils is very poor and under strongly to moderately acidic soils, the plant growth and development is affected to a great extent. The crops grown on such problematic soils do not give remunerative returns. One of the most important and practically feasible management practices is the use of Lime and Liming Materials to ameliorate the soil acidity. The addition of lime raises the soil pH, thereby eliminating most major problems of acid soils. Cultivation of acid tolerant crops/varieties is also possible option.

Lime requirement (LR): Is defined as the amount of liming material that must be added to raise the pH to some desired value (pH 6.0 to 7.0). There are a number of liming materials available in the market and may be used according to their availability, convenience and economy. Some of the important liming materials are;

1. Lime oxide (CaO)- Burnt lime/quick lime or oxide of lime. Caustic in nature and difficult to handle.
2. Lime hydroxide {$Ca(OH)_2$}- hydrated lime or slaked lime. Caustic in nature and difficult to handle.
3. Lime carbonate ($CaCO_3$)- ground lime stone occurs as Boglime or marl, Calcite, Oysters shells and precipitate carbonates. It is most commonly used and it is largely sold in the market as pulverized or ground lime stone, called agricultural lime.
4. Basic slag: Byproduct of Iron industry.
5. Dolomite ($CaMgCO_3$)

Liming reaction in soils

$H^+ [Soil] + CaCO_3$ ——— $Ca^{2+} [Soil) + CO_2 + H_2O$

or

$CaCO_3 + H_2CO_3$ ——— $Ca(HCO_3)_2$

$H^+ [Soil] + Ca(HCO_3)2$ ——— $[Soil]Ca2^+ + 2 H2O + 2CO2$

Benefits of liming on acidic soils

1. Direct benefits
   a. Addition of lime raises the soil pH to a desired level.
   b. It eliminates major problems of Fe, Al, Mn and H toxicities.
2. Indirect benefits
   a. The nutrients like Ca^{2+} and Mg^{2+} are added to the soil if lime is dolomite ($Ca, MgCO_3$)
   b. The raised pH also reduces excess soluble Mn and Fe by forming insoluble hydroxides.

3. Phosphorus availability is improved or increased because of liming precipitates Fe and Al in other forms (P is fixed as Fe and $AlPO_4$).
4. It makes K^+ more efficient in plant nutrition.
5. It increases the availability of N by creating a more favorable environment for microbes.
6. Bacterial activity is enhanced and help in control of certain plant pathogens like club rot disease of cole crops.

Calcareous soil, their formation, characteristics & management

Calcareous soils are those that contain enough free calcium carbonate ($CaCO_3$) and give effervescence visibly releasing CO_2 gas when treated with dilute 0.1 N hydrochloric acid. The pH of calcareous soil is > 7.0 and also regarded as an alkaline (basic) soil.

Formation

The soil is formed largely by the weathering of calcareous rocks and fossil shell beds like varieties of chalk, marl and lime stone and frequently a large amount of phosphates. Soils are often very fertile, thin and dry. They are found in large part of arid and semiarid regions, which may prove very fertile when sufficient moisture for crops is available. Soils also can become calcareous through long term irrigation with water contains small amounts of dissolved $CaCO_3$ that can accumulate with time. Calcareous soils can contain from 3% to >25% $CaCO_3$ by weight with pH values with a range of 7.6 to 8.3.

Management of calcareous soil

Fertilizer management in calcareous soils is different from that of non calcareous soils because of the effect of soil pH on soil nutrient availability and chemical reactions that affect the loss or fixation of some nutrients. The presence of $CaCO_3$ directly or indirectly affects the chemistry and availability of nitrogen (N) Phosphorus (P), Magnesium (Mg), Potassium (K), Manganese (Mn), Zinc (Zn) and iron (Fe). The availability of copper (Cu) also is affected. Application of acid forming fertilizers such as ammonium sulphate and urea fertilizers, sulphur compounds, organic manures and green manures is considered as effective measures to reduce the pH of soil to neutral pH value.

Chapter 7

Salt Affected Soils Their Formation and Management

1. Salt affected soils;
2. Classification of salt affected soils;
3. Formation of salt affected soils;

A. Saline soils & their management

a. Characteristics of saline soil;
b. Formation/Development of saline soils;
c. How to diagnose salinity problems in the field?
d. Soil salinity classes and crop growth;
e. Effect of salinity on crop growth;
f. Methods of Reclamation and Management Saline soils.
   I. Physical methods
   II. Agronomic and cultural methods

B. Alkali soils and their management

a. Characteristics of Alkali soils;
b. Formation of Alkali soil and their classification;
c. How to diagnose Alkali Soils in the field?
d. Problems of alkali soils and plant growth;
e. Reclamation and management of alkali soils.
   1. Use of Chemical/Industrial amendments
   2. Organic amendments
   3. Cultural practices.

4. Water management
5. Method of application of amendments
6. Sodicity classification according to the FAO.

1.Salt affected soils

All soils contain some amounts of soluble salts. Many of these salts act as a source of essential nutrients for the healthy growth of plants. However, when the quantity of the salts in the soil exceeds a particular limit, the growth, yields and/or quality of most crops is adversely affected depending upon the kind and amount of salt present. According to Central Salinity Research Institute, Karnal, the area of salt affected soils (Soil salinity plus alkalinity) in India is 7421 thousand hectares (approximately).

Soil that contains excess amount of soluble salts such as chlorides, sulphates, carbonates and bicarbonates of calcium, magnesium and sodium so as to impair productivity of crops is called salt affected soil.

Classification of salt affected soils

Indian system of classification: Classified into 2 major types of salt affected soils based on pH, EC and ESP.

Soil Characteristics	Saline soils	Alkali soils
pH(s)	< 8.2	> 8.2
ESP	< 15	> 15
EC(e)	>4 ds/m	Variable, mostly <4dS/m
Nature of Soluble salts	Neutral mostly Cl^- & SO_4^{2-}, HCO_3^- may be present but CO_3^{2-} is absent	Capable of alkaline hydrolysis, preponderance of HCO_3^- & CO_3^{2-} of Na^+.

Formation of salt affected soils

Nature and amount of soluble salts play an important role in the formation salt affected soils. Soluble salts are those inorganic chemicals that are more soluble than gypsum. Soluble salts in soils are mainly composed of the cations like sodium (Na^+), Calcium (Ca^{2+}) and magnesium (Mg^{2+}) and potassium (K^+) and ammonium (NH_4^+) in small amount. The anions like chlorides (Cl^-). Sulphates (SO_4^{2-}) carbonates (CO_3^{2-}) and bicarbonates (HCO_3^-). The salt formation is therefore expected to result from a combination of these groups such as NaCl, Na_2SO_4, $MgSO_4$, $CaCl_2$, $CaSO_4$, Na_2CO_3, $NaHCO_3$, and $MgCO_3$ and $CaCO_3$.

Salt affected soils formed mainly in 2 phases or processes

1. **Salinization process:** Process of accumulation of soluble salts in the soil results in the formation of saline soil.
2. **Alkalization process:** Process of accumulation of exchangeable sodium in the soil & formation of sodic soils is termed as alkalization.

A. Saline soils & their management

Saline soils

They are defined as soils containing appreciable amounts of soluble salts like chlorides, sulphates of Ca, Mg, K, and Na *etc.* that interfere with plant growth.

Characteristics of saline soil

The EC of saturation paste extract of these soils is > 4dSm^{-1}, pH of the saturation paste is < 8.2 and the ESP is < 15. They are also known as white alkali soils or solon chalks. These soils lack structural B horizon and contain very little organic matter (<1%). Soluble salts mostly consist of chlorides and sulphates of Na^+, Ca^{2+} and Mg^{2+}. Bicarbonates may or may not be present. However carbonates are generally absent.

Formation of saline soils

Saline soils are formed whenever climate, soil and hydrological conditions favour accumulation of soluble salts in the root zone. Arid and semi-arid regions of low rain fall and high temperature leads to less leaching, more evaporation of water and less transport of soluble salts. Irrigation with salt-laden underground water and their indiscriminate use under inadequate drainage creates more salinity problems. Excessive use of chemical fertilizers may also leads to form saline saline soils.

Diagnosing salinity problems in the field

In the field, saline soils can be identified by:

- Presence of white crust layer of salts on the surface of soils in a dry region. It is a mixture of NaCl and Na_2SO_4 salts.
- Good physical conditions and high permeability of soil.
- Natural vegetation consisting of mainly small bushes and some halophytes such as *Cressacretica, Cyperusrotudus, Chlorisgayana etc.*
- In cultivated fields patchy and stunted growth of crops.
- Wilting sign due to water stress in plants even when the soil apparently contains enough water due to plasmolysis.

- A thin soil crust that may prevent emergence of seedlings of sensitive crops.

Soil salinity classes and crop growth

Sl. No	Soil salinity class	EC(e).dSm-1	Effect on crop growth
1.	Non saline	<2	Salinity effects negligible
2.	Slightly saline	2- 4	Yields of sensitive crops may be restricted.
3.	Moderately saline	4 - 8	Yield of many crops restricted.
4.	Strongly saline	8 - 16	Only tolerant crops yield satisfactory.
5.	Very strongly saline	>16	Only a few very tolerant crops yield is satisfactory.

Equations for relating EC with OP and TDS

Conversions

Total cations (m eq/L)=10 X EC(dS/m) at 25°C

Salt Concentration(mg/L)= 640 X EC(dS/m) at 25°C

Osmotic Pressure= 0.36 X EC (dS/m) at 25°C

$$\%\ \text{Salt in solution} = 0.064 \times EC\ (dS/m)\ \text{at}\ 25^\circ C \times \frac{\text{Saturation percentage(SP)}}{100}$$

Effect of salinity on crop growth

Excess amount of soluble salts affects the plant in the following ways;

1. **Water availability to plants decreases**: with increase in salt concentration of the soil the osmotic pressure of the soil solution increases and plants are not able to extract water as easily as they can from the soil. Therefore, EC (e) of the soil increases the water becomes less available to the plant even though the soil may contain water and appears quite moist. Due to high osmotic pressure water comes out of the plant which is called reverse osmosis or plasmolysis. The plant show wilting symptoms which referred as physiological drought.

2. **Specific Toxic effect of ions**: Higher concentration of individual ions in the root environment or in the plant may prove toxic to the plant or may retard the uptake (absorption) and metabolism of the essential plant nutrients and thus affect the normal growth of the plant.

Antagonism between Cl^- and $H_2PO_4^-$, Cl^- and NO_3^-, Cl^- and SO_4^{2-} and Na^+ and K^+ may disturb the normal nutrition of plants.

Methods of reclamation or Management of saline soils

I. Physical methods

- **Scraping:** The salts accumulated on the surface can be removed by mechanical means. This is the simplest & most economical way to reclaim saline soils if the area is very small e.g. small garden lawn or a patch in a field. This improves plant growth only temporarily as the salts accumulate again & again.
- **Flushing:** Washing of surface salts by flushing water. This is especially practicable for soils having a crust & low permeability. However this is not sound method of practice.
- **Leaching:** leaching with good quality water, irrigation or rain is the only practical way to remove excess salts from the soil. It is effective if drainage facilities are available, as this will lower the water table & remove the salts by draining the salt rich effluent.

Leaching requirement: The amount of water needed to remove the excess soluble salts from the saline soils is called leaching requirement or The fraction of the irrigation water that must be leached through the root zone or soil profile to control soil salinity at any specific level (Salt balance).

$$\text{Leaching Requirement (LR)} = \frac{EC_{iw}}{EC_{dw}} \times 100 \quad \text{or} \quad \frac{D_{dw}}{D_{iw}} \times 100$$

Where,

EC = Electrical Conductivity in dSm^{-1},

iw = EC of Irrigation water in dSm^{-1},

dw = EC of Drainage water in dSm^{-1},

D_{dw} = Depth of drainage water in inches and

D_{iw} = Depth of irrigation water in inches

II. Agronomic and cultural methods: In areas where only saline irrigation water is available or when shallow saline water table prevails and soil permeability is low, the following cultural practices are adopted.

1. Selection of crops and crop rotations

On the basis of crop tolerance to quality of irrigation water or soil salinity the crops can be classified in four groups, *viz.*

- **Highly tolerant crops:** Barley, Sugar beet, datepalm.
- **Tolerant:** Tapioca, mustard, coconut, spinach, amaranthus, pomegranate, guava, ber.
- **Semi-tolerant:** Ashgourd, bitttergourd, brinjal, cabbage, cluster bean, pea, lady's finger, muskmelon, onion, potato, Indian bean, sweet potato, tomato, turnip, water melon.
- **Sensitive:** Radish, carrot, coriander, cumin, mint, grape, sweet orange.

2. Method of raising plants

a. Crop should be raised by transplanting seedlings (especially vegetables, flowers, fruit trees) than germinating the seeds.

b. Wild root stocks grafted with a good quality but salinity sensitive scion. (Mango, citrus, Guava and ornamental plants like rose).

3. Irrigation practices

a. Method of water application-follow furrow or drip irrigation, sub surface irrigation systems and sprinkler irrigation.

b. Frequency of irrigation- irrigation more often (frequent) can maintain better water availability & decrease the salinity should not too much irrigations.

4. Use of mulching materials to prevent evaporation losses like straws, plastic sheets *etc*.

Chapter 8

Alkali Soils and Their Management

Alkali soil

Alkali soils are those that contain measurable amounts of soluble salts mostly as carbonates and bicarbonates of sodium.

Characteristics of alkali soils

These soils have pHs> 8.2 and ESP > 15. The EC_e of alkali soils is variable but normally <4 dSm^{-1}. Neutral salts like NaCl and Na_2SO_4 are present in very small quantities. The alkali soils are also known as sodic or solonetz. They are also called as Black alkali soils. It is mainly because of high pH and Na_2CO_3 the finely decomposed organic matter is dissolved along with the water that imparts a dark black or brown colour to the soil.

Formation of alkali soils

1. Sodication due to repeated cycles of wetting & drying.
2. Sodication due to shallow brackish ground waters.
3. Sodication due to the use of high RSC water.
4. Sodication due to the use of saline irrigation waters low in Cl: SO_4 ratio.

In contract to saline soils the saturation extract of alkali soils contains measurable amounts of CO_3^{2-} and HCO^{3-}; Cl^- and SO_4^{2-} may also be present in appreciable amounts. Among the cations, sodium (Na+) is the dominant one followed by Mg^{2+} and Ca^{2+}.

Table: Sodicity classification according to the FAO

Class	ESP	Description
a	<10	Sodic free soil
b	10-20	Slightly sodic
c	20-30	Moderately sodic
d	30-50	Straggly sodic
e	>50	Very sodic

Characteristics of sodic soils

- There will not be any natural vegetation except some very hardy grasses.
- When wet, alkali soils turn black because of the humic acid fraction of organic matter, which is dissolved by Na_2CO_3at high pH.
- They are very slippery and soft when wet but very hard when dry.
- Upon drying deep cracks, 1-2 cm wide develops in the soil which closes when wetted.
- The surface soil develops a hard crust with typical convex surfaces.
- Water movement is restricted due to the sealing effect of Na- clays. A few centimeters below the surface may therefore be almost saturated with water while the surface is dry & vice versa.
- Runoff water is always turbid because of the bad physical conditions.
- Clay pan formation or kankar pan formation at variable depths.

Problems of alkali soils on plant growth

- Poor water and air permeability as a result of high dispersion of soil aggregates and clay particles.
- Low availability of water due to poor conductance from the lower soil layers.
- Seedling emergence & germination is greatly affected due to hard crust surface.
- Seed bed preparation is very difficult because it does not come to proper moisture conditions.
- Deficiency of Ca and there is an antagonistic effect on K and Ca nutrition.
- Toxic concentration of $HCO_3^-$and CO_3^{2-} ions.
- Decreased solubility of Fe & Zn.
- Increased solubility and accumulation of toxic elements such as F, Se & Mo in plants that may affect crop yield.

Reclamation or management of alkali soils

I. Use of Amendments: For successful crops in alkali soils ESP of the soil must be lowered which can be achieved by application of amendments.

Gypsum ($CaSO_4 2H_2O$) is the most commonly used amendment for reclamation. Elemental S and pyrite are also used.

Gypsum requirement (GR)

The amount of gypsum needed for removal of exchangeable sodium from alkali soil and also to decrease the soil pH to a desired level is called gypsum requirement. 1 milli equivalent of Ca^{2+} is required to replace 1 milli equivalent of Na^+ ion from the exchange complex of sodic soil.

Method of application of amendments

Amendment should normally be broadcasted and incorporated in the surface 10 cm soil. Deeper depth of mixing of gypsum becomes inactivated and thereby reduces the effectiveness of the applied gypsum and yield of crops. To ensure proper dissolution of gypsum and leaching of replaced exchangeable Na^+, the water is to be ponded/stagnated continuously on the soil for a period of 8-10 days (80% of gypsum gets dissolved). It brings down soil exchangeable sodium percentage (ESP) to a desired level before sowing. When pyrites or elemental sulphur are used as an amendment, then a cycle of wetting and drying is needed to maintain proper moisture and aerobic conditions to ensure maximum oxidation and higher production of H_2SO_4.

Reactions of amendments in soil

If Gypsum used

$Na_2CO_3 + CaSO_4 CaCO_3 + Na_2SO_4$

$2NaHCO_3 + CaSO_4 Ca(HCO_3)_2{}^+ Na_2SO_4$.

Or

$2Na^{+-}\text{-Soil}^+ + CaSO_4 Ca^{2+} - \text{Soil} + Na_2SO_4$.

If Elemental sulphur used

$2S + 3O_2 2 SO_3$ (Microbial oxidation)

$SO_3 + H_2 0 H_2SO_4$.

$H_2SO_4 + CaCO_3 CaSO_4 + CO_2 + H_2O$

$2Na X + CaSO_4 Ca\text{-}X_2 + Na_2SO_4$

a. Organic amendments: Addition of crop residues and other organic materials have got a tremendous influence on maintaining soil physical, chemical and biological properties of these soils. Organic amendments like straw, rice husk. Ground nut and safflower hulls, FYM, compost, poultry droppings, green manures, tree leaves and saw dust. They Produce CO_2 (carbonic acid) and organic acids, they increase the solubility of calcite and lower the ESP.

b. Industrial by products as amendments: Phospho-gypsum, Press mud, Molasses, Acid wash and Effluent from milk plants. They are used to provide soluble Ca directly/indirectly by dissolving soil lime.

c. Cultural practices

- *Land leveling and shaping:* To ensure proper water management and uniform leaching of salts the field should be leveled properly.
- *Maintain Plant population:* Crop stand in alkali soils can be improved by increasing the seed rate and reducing the planting distance.
- Age of seedlings - planting old seedlings has proved to be beneficial.
- *Green manuring:* Increases organic matter content, there will be releasing of CO_2 and formation carbonic acid. This lowers the pH and it enhances solubility of native $CaCO_3$ and adds a considerable amount of plant nutrients in the soil.

  Ex: Sesbania, diancha, subabul
- *Continuous cropping:* the land should be continuously cropped to keep the downward movement of replaced Na & soluble salts. Growing of rice crop, along with other practices has been observed to enhance the reclamation of alkali soils.

d. Water management

1. **Drainage:** Alkali soils have got low infiltration rate and all the rain water accumulates to create surface water logging. This results oxygen stress. So, to avoid this problem provide surface and subsurface drainage system in the field.

2. **Good irrigation management:** Normally surface method of irrigation such as furrow or basin type flood method is used for alkali soils. However the sprinkler method could be promising because of its ability to supply water uniformly and in small quantities/

PART - II : NUTRIENT MANAGEMENT

In this portion the principles of nutrient management are described. The plant uptake poses demand for nutrients and the soil supplies them. These (plant uptake and nutrient availability) can be quantified and nutrients can be well managed with economic sense to meet their demand. These principles are elucidated.

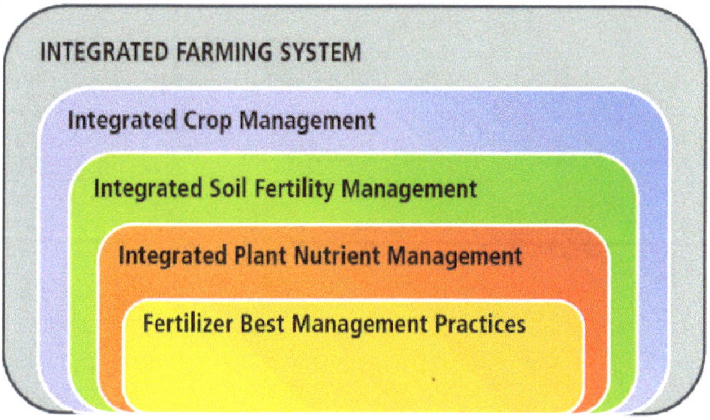

Fig. 1: Different Aspects of Nutrient Management are a part of Integrated Farming System

Important factors that need to be considered when managing nutrients include

(a) The application of nutrients considering the achievable optimum yields and, in some cases, crop quality;

(b) The management, application, and timing of nutrients using a budget based on all sources and sinks active at the site; and

(c) The management of soil, water, and crop to minimize the off-site transport of nutrients from nutrient leaching out of the root zone, surface runoff, and volatilization (or other gas exchanges).

There can be potential interactions because of differences in nutrient pathways and dynamics. For instance, practices that reduce the off-site surface transport of a given nutrient may increase the leaching losses of other nutrients. These complex dynamics present nutrient managers the difficult task of achieve the best balance for maximizing profit while contributing to the conservation of our biosphere.

Chapter 9

Integrated Plant Nutrient Management (IPNM)

Definitions of IPNM

Concept of IPNM

Main objectives of INM

Components of IPNM

I. Use of inorganic fertilizers

II. Use of organic manures/ materials

III. Use of biological sources/biofertilizers

IV. Maintaining the physical properties of soil

V. Management of problematic soils

VI. Better/Judicious water management practices

Definitions

IPNM is the intelligent and combined use of inorganic, organic and biological resources so as to sustain optimum yields improve or maintain the soil chemical and physical properties and provide crop nutrition packages which are technically sound, economically attractive, practically feasible and environmentally safe. The principal aim of the integrated approach is to utilize all the possible sources of plant nutrition in a judicious & efficient manner.

Concept of IPNM

The basic concept of IPNM is the promotion & maintenance of soil fertility for sustaining crop productivity through optimizing all possible sources like organic, inorganic and biological in an integrated manner appropriate to each farming situation. Improvement of soil fertility and productivity on sustainable basis through appropriate use of fertilizers and organic manures and their scientific management for optimum yield, growth of different cropping systems in a specific agro ecological conditions.

Main objectives of IPNM or INM

1. To reduce the dependence on chemical fertilizers.
2. To maintain productivity on sustainable basis without affecting soil health.
3. To conserve locally available resources & utilize them judiciously.
4. To reduce the gap between nutrients used & nutrients harvested by the crop.
5. To improve physical, chemical & biological properties of soil.
6. To make soil healthy by providing balanced nutrients through different nutrient sources.
7. To overcome or reduce the ill effects of continuous use of only inorganic chemical fertilizers.
8. To improve economical status of farmers.
9. To increase the fertilizer use efficiency (FUE).

Components of IPNM

IPNM mainly emphasizes the integrated use of all the essential nutrients from different sources like chemical fertilizers, organic manures, green manures, bio-fertilizers, legume crops, locally available plant resources in a balanced proportion for sustainable soil health and productivity.

I. Use of inorganic fertilizers: They are very important for sustaining and increasing food production. Different kinds of fertilizers are commercially available in the market for all the major and micronutrients. However, they are costly inputs and their excessive use may deteriorate the soil quality and food quality. Hence, there is a need to improve their use efficiency through efficient and balanced fertilizer management and essentially follow the four R's formula for judicious and effective nutrient/fertilizer management. They are:

- Right type of fertilizers.
- Right dose of fertilizers.
- Right method of application.
- Right time of application.

Components of 4R Nutrient Stewardship Programme

Right source	Provide plant-available forms, and a ba lanced supply of all essential nutrients. Take advantage of various formulations that offer improved efficiency and reduce environmental consequences.
Right rate	Ensure an adequate supply of all essential nutrients to meet plant demand.
Right time	Manage nutrient applications to match the interactions of crop uptake, soil supply, environmental risks, and field operation logistics.
Right place	Consider root-soil dynamics and nutrient movement, and manage spatial variability within the field to meet site-specific crop needs and minimize potential losses from the field.

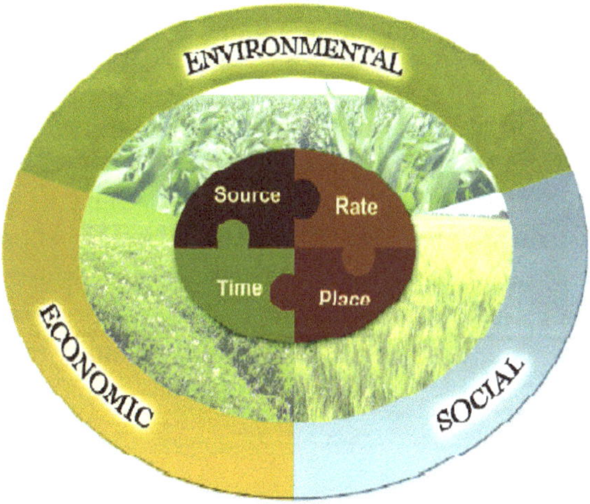

4R Stewardship Concept: Locking of Social economic and Environmental Goals

II. Use of organic manures/materials: Due to intensive cultivation of soil and less organic manure application, the soils are low in organic matter status. A decrease in soil organic matter results in compact soil, poor aeration and low infiltration and water holding capacity and also low fertility status. The organic matter status in soils can be improved and maintained by constant addition of organic manures such as FYM, compost, green manures, poultry manures, vermicompost, oilcakes *etc*. Organic matter is good source of macro and micro nutrients and more over improves physical, chemical and biological properties of soil.

III. Use of biological sources/biofertilizers: Biofertilizers are cultures of micro organisms (bacteria, fungi, algae). Their use benefits the soil and plants growth by providing N & P and also brings about the rapid mineralization of organic materials in soils. They are capable fixing N, solubilizing and mobilizing the phosphorus and mineralizing organic matter in soil. Their incorporation improves the physical and biological properties of soils.

IV. Maintaining the physical properties of soil: Physical properties such as soil aggregation, soil texture, structure, aeration, water holding capacity (WHC), infiltration rate, etc., should be maintained regularly through better cultivation practices and organic manure applications to maintain soil fertility & nutrient availability.

V. Management of problematic soils: Problematic soils such as acid soils, saline and alkaline soils, waterlogged soils are known to decrease the productivity of the soil. Acid soil having the problems like toxicities of Iron, Mn, Al, deficiency of P & Mo. Similarly, saline and alkali soils showing the deficiency of Fe, Mn, Zn and Cu and also toxicities of Mo. These soils should be regularly managed and reclaimed through the application of soil amendments such as lime for acid soil, gypsum for alkali soils and other organic and inorganic materials based on soil test results. It helps to improve soil fertility and productivity and sustain the yield.

VI. Better/Judicious water management practices: Plants absorb the nutrients from the soil only in a dissolved state and sufficient moisture is therefore required for utilizing the nutrients of the soil. Management of moisture in the soil by improved and modern irrigation techniques like drip or sprinkler or basin where the rainfall is low and draining the soil where it is subjected to stagnation of water helps to increase water and nutrient availability to the crops.

Chapter 10

Nutrient Use Efficiency (NUE) and Management

The objective of nutrient use is to increase the overall performance of cropping systems by providing economically optimum nourishment to the crop while minimizing nutrient losses from the field and supporting agricultural system sustainability through contributions to soil fertility or other soil quality components. NUE addresses some but not all aspects of that performance. The most valuable NUE improvements are those contributing most to overall cropping system performance. Therefore, management practices that improve NUE without reducing productivity or the potential for future productivity increases are likely to be most valuable. An excellent review of NUE measurements and calculations was written by Dobermann (2007). Table is a summary of common NUE terms, as defined by Dobermann, along with their applications and limitations. The primary question addressed by each term and the most typical use of the term are also listed below.

Effect of low NUE

Low NUE (<50% in N) challenges sustainable intensification of farming soils in the face of exhausted possibility to amplify production and remain food secure. Saving land by increasing productivity/unit area remains the only option to feed the burgeoning population. Development of non-sustainable intensification is, thus, a consequence of man's quest to raise productivity without imposing holistic management. Holistic management, includes balancing yield maximization with minimization of disruption to the ecological services. Exclusive focus on yield pushing management practices, in particular, destabilize, natural nutrient cycle. 'Flux of nutrients across physical environment (soil, water and air), living organisms and back to physical environment' what constitutes a natural N (also other nutrients) cycle. Overdevelopment of native nutrient harvests (or soil mining) is a typical offshoot of disturbed native N cycle. Loss of useful soil

Table: Common NUE terms and their application (after Dobermann, 2007).

Term	Calculation*	Question addressed	Typical use
Partial factor productivity	PFP = Y/F	How productive is this cropping system in comparison to its nutrient input?	As a long-term indicator of trends.
Agronomic efficiency**	AE = (Y-Y0)/F	How much productivity improvement was gained by use of nutrient input?	As a short-term indicator of the impact of applied nutrients on productivity. Also used as input data for nutrient recommendations based on omission plot yields.
Partial nutrient balance	PNB = UH/F	How much nutrient is being taken out of the system in relation to how much is applied?	As a long-term indicator of trends; most useful when combined with soil fertility information.
Apparent reco-very efficiency by difference	RE = (U-U0)/F	How much of the nutrient applied did the plant take up?	As an indicator of the potential for nutrient loss from the cropping system and to access the efficiency of management practices.
Internal utilization effi-ciency	IE = Y/U	What is the ability of the plant to transform nutrients acquired from all sources into economic yield (grain, etc.)?	To evaluate genotypes in breeding programs; values of 30-90 are common for N in cereals and 55-65 considered optimal.
Physiological efficiency	PE = (Y-Y0)/ (U-U0)	What is the ability of the plant to transform nutrients acquired from the source applied into economic yield?	Research evaluating NUE among cultivars and other cultural prac-tices; values of 40-60 are common.

* Y = yield of harvested portion of crop with nutrient applied; Y0 = yield with not nutrient applied; F = amount of nutrient applied; UH = nutrient content of harvested portion of the crop; U = total nutrient uptake in aboveground crop biomass with nutrient applied; U0 = nutrient uptake in aboveground crop biomass with no nutrient applied; Units are not shown in the table since the expressions are ratios on a mass basis and are therefore unitless in their standard form. P and K can either be expressed on an elemental basis (most common in scientific literature) or on an oxide basis as P2O5 or K2O (most common within industry).

biology, damage to soil health/quality and unprecedented changes in climatic patterns are other negative end products. Direct Impact of fertilizer-N on useful soil biology is transitory. Likewise, impairment of soil quality, in terms of chemical reactivity or physical attributes, in general, is offset by the inherent soil buffering capacity. Nevertheless, it is the poor NUE propelled global warming due to N_2O gas emissions or eutrophication that is the source of decaying soil health. Besides, poor NUE necessitates higher consumption and production of urea-N. Each ton of elevated production of urea-N costs 47 barrels of crude – estimated IFDC. Then expenditure of every 100 barrels of crude pumps 47 tons CO_2 in to the atmosphere. Poor NUE is a double whammy triggering global warming. It manifests in terms of heightened temperatures mediated reduction in labile/ active pool (light fraction of C) of soil organic matter. Since with climate change extreme rain events are becoming frequent, even sustainability of stable carbon fraction (heavy fraction of C) gets weakened. Whether it is the plunder of active or recalcitrant carbon fraction, needless it is to mention, soil health suffers in all aspects.

NUE management

It depends on holistic fertilizer management, It is inevitable to:

(i) Make soil test and crop need based applications that not only equal crop removals, but also balances fertilizer treatment in proportions that adequately ameliorate all deficient nutrients.

(ii) Adopt efficiency enhancing fertilizer methods, times, sources and doses and

(iii) Include supplementary treatment with indigenous sources and resources. Besides, these basic elements of efficient fertilizer management.

(iv) It is inescapable to exclude practices like standard/precise agricultural methods and smart use of pesticides, water and energy.

(v) Above all, it mandates to prepare expert human resource for transferring all-inclusive information on a right management package. Working in a participatory mode with all stakeholders will hasten the understanding prompting lasting adoption of holistic fertilizer management by the client farmers.

(vi) Despite these well-meaning accomplishments, it is a paradox that level of NUE today remains what it was 50 years back. New developments like site specific nutrient management, for example, (software Nutrient Expert or green sensors Green Seeker) or leaf color chart aided fertilizer dressings and role of sequencing fertilizer application with reference to irrigation should be adopted.

with these practices the NUE of N can be realized up to 80 percent. Use efficiency of P and K is not a substantive issue as they are held by the soil colloids. For P adoption of P optimization techniques can increase its NUE. (See also Fertilizer Use Efficiency and Management).

Leaf Colour Chart (LCC) for rice (IRRI)

SPAD Chlorophyll meter for greenness measurement

Greenseeker -on -the go sensing system

(See colour version on page 144)

Chapter 11

Fertilizer Use Efficiency (FUE) and Management

Fertilizers are applied to supplement nutrient requirement of the crop. It should not be looked as a substitute to organic sources. After determination of nutrient requirement of a crop fora given yield and contribution of nutrients from different sources, articularly, from the soil source, it is necessary to supplement the balance from the inorganic sources. These are determined by field experimentation supplemented by pot-culture, laboratory and green housestudies, if necessary. When a fertilizer is applied all of its nutrient (s) are not absorbed by the crop. The interactions between soil-crop-season and other factors are quite significant. Only a fraction of the nutrient(s) is utilized by the crop.

Efficiency in any system is an expression of obtainable output with the addition of unit amount of input. The ratio of energy intake and energy of the produced biomass *i.e.* of input and output is called ecological efficiency. This can be studied at any trophic level. Fertilizer use efficiency is the output of any crop per unit of the nutrient applied under a specified set of soil and climatic conditions.

The NUE/FUE can be expressed in several ways. Mosier *et al.* (2004) described four agronomic indices to describe NUE: partial factor productivity (PFP, kg crop yield per kg input applied); agronomic efficiency (AE, kg crop yield increase per kg nutrient applied); apparent recovery efficiency (RE, kg nutrient taken up per kg nutrient applied); and physiological efficiency (PE, kg yield increase per kg nutrient taken up). Crop removal efficiency (removal of nutrient in harvested crop as % of nutrient applied) is also commonly used to explain nutrient efficiency. According to Barker (1977), fertilizer efficiency is increase in yield per unit of fertilizer nutrient applied; the agronomic efficiency.

The partial factor productivity (PFP) from applied nutrients is a useful measure of nutrient use efficiency because it provides an integrative index that quantifies total economic output relative to utilization of all nutrient resources in the system.

For a given yield level, optimum factor productivity from applied nutrients is achieved when the use of indigenous soil nutrients is maximum and the efficiency of applied nutrients in producing economic yield is high. In the long term, specification of optimum factor productivity must also consider nutrient balance so that the depletion of nutrient stocks below critical threshold levels does not lead to increased requirements for applied nutrients to maintain yield levels.

Nutrient Recovery (NR) refers to the actual amount of Nutrient taken up from the fertilizers.

$$\text{FUE in terms of NR (\%) through fertilizer} = \frac{\text{Quantity of the nutrient of fertilizer taken up by the crop}}{\text{Quantity of nutrient applied}} \times 100$$

NR could be best known through isotopic studies. For practical purposes, however, it is worked out as follows:

$$\text{NR (\%)} = \frac{\text{Nutrient uptake in fertilized plots - Nutrient uptake in unfertilized plots}}{\text{Amount of nutrient added}} \times 100$$

The NR exhibits a diminishing trend with the increasing amounts of nutrients.

Agronomists usually express the efficiency of fertilizer nutrient in kg of grain yield produced kg-1 of applied nutrient, Agronomic efficiency (AE). It refers to the additional produce obtained in kg per kg of an applied nutrient. It is calculated as follows:

$$\text{AE} = \frac{\text{Yield from fertilized plot in kg} - \text{yield without fertilizer in kg}}{\text{Amount of fertilizer nutrient applied in kg}} \times 100$$

The AE like NR generally decreases with the increase in nutrient supply.

Physiologists, define the efficiency of nutrient utilization in kg of grain yield produced kg-1 of nutrient absorbed. These two efficiencies (Agronomic and physiological) can be related by introducing, percentage of nutrient recovery (a third parameter worked out as above).

Efficiency of fertilizer nutrient (kg yield kg-1 nutrient applied)	=	Percentage of nutrient recovery (kg absorbed nutrient kg^{-1} applied nutrient)	x	Efficiency of utilization (kg yield kg^{-1} nutrient absorbed)

The percentage nitrogen recovery varies with soil properties; methods, amounts and timing of fertilizer application and other management practices. It usually ranges from 30 to 50 per cent in the tropics.

Techniques for increasing fertilizer use efficiency (Best fertilizer management practices)

It is reasonable to believe that at least one half of the observed yield gap (up to 3 tons/ha) can be assigned to inefficient use of fertilizers.To increase the fertilizer use efficiency the nutrient must be available at the time of its requirement by the crop, in right form and quantity. On application there occur certain inevitable/ evitable losses of nutrients that reduce the efficiency. The losses are due to: (i) leaching, (ii) volatilization, (iii) immobilization, (iv) chemical reaction between various components in the mixture, (v) change in capacity to supply nutrients, and (vi) unfavourable effects associated with fertilizer application.

Each component of loss can be reduced to a great extent by management of the soil- fertilizer-crop system. This requires knowledge and experience on (i) how much of the fertilizer to be applied, (ii) what/which (type of fertilizer) to be applied, (iii) when to be applied (time of application), (iv) how (method of application), (v) where (placement of fertilizer) and (vi) other considerations (cost, availability of fertilizer, labour, ease of application, awareness on benefits of fertilizer use, etc.).

How much

Inorganic source is a supplement to other sources of nutrients. Among other sources, the most important one is soil source. Availability of nutrients from soil and fertilizer sources can be estimated from field experiments involving response to fertilizers and tracer techniques (using radio-active isotopes).

What and which (Type of fertilizer)

Fertilizers vary with respect to their solubility besides their grade. Choice of fertilizer is location specific and needs to be found out by field experimentation. The choice is more with respect to nitrogen and phosphatic fertilizers than for potassic. Studies on crop response is also more for N than for P or K fertilizers because leaching loss is more in nitrogenous fertilizer and its residual effect is nil or negligible. In case of P, its indirect, residual and cumulative effects are more important.

Nitrogen in form of NO_3- is subject to more leaching. Leaching loss is also more in wet *(kharif)* than in summer and in sandy soils than in clayey soils. Losses can be minimized by choosing suitable time and method of application.

When to apply

It necessarily means time of application. The objective of time of application is to get maximum benefit from the fertilizer nutrient. If the nutrient is applied too earlier than the time of requirement, it is lost in different ways or is absorbed more than required. If applied late it is either not absorbed or if absorbed not utilized for the purpose and only gets accumulated in plant parts. Some amendments need to be applied before commencement of crop season so that it reacts well with the soil and becomes available to the crop after sowing/ planting (Example-application of press mud, other liming materials, bone-meal, Ferrochrome slags-bhusakti, etc).

Where to apply (placements)

The objective of placement of fertilizer is to make the nutrient available easily to the crop. It should be near to the roots. Application may be surface broadcast, at furrow bottom, placed deep at or slightly below the root zone, top dressed, side dressed or to foliage. This depends on type of crop, rooting pattern, feeding area and ease of application. The choice of method of application depends on soil-crop-fertilizer interaction too.

Other considerations

- Proper control of pests and diseases is must for realizing maximum effectiveness from fertilizers.
- Weeds, if not controlled effectively particularly during early stages (7-21 days) of crop growth in kharif season, take away about 25 to 30 per cent of the applied plant nutrients. Therefore, the weed control, particularly during early stages of crop growth is essential.
- When the soils are acidic or saline or alkali, appropriate amendments viz. lime, gypsum *etc*. should be applied before using fertilizers. In alkali soils 3 to 5 tonnes of gypsum per acre (8 to 12 t/ha) should be applied broadcast only once and mixed with the top 10 cm of the soil layer.
- Rock phosphates can be profitably used in acid soils and in low land rice and legumes.
- Deficiency of Zn is becoming increasingly widespread. In such cases 25-50 kg zinc sulphate should be applied through soil as basal application. If symptoms of zinc deficiency appear in standing crop, it should be sprayed with 0.3 to 0.5 per cent solution of zinc sulphate mixed with 0.3 percent solution of lime for quick recovery.
- Under adverse soil and climatic conditions *e.g.* light permeable soils, rainfed conditions or where the crop is grown under deep standing water, application of fertilizers (particularly N) through foliage along with insecticides and pesticides (if needed will lead to higher utilization efficiency by plants).

- For rice crop, wherever possible, mix urea with available nitrification inhibitors such as neem cake and karanj cake (1 kg of cake blend with 5 kg of urea). This will reduce N losses from the soil. Curing urea with soil for top-dressing reduces N loss. Mixing one part of urea with 5 to 10 parts of moist soil thoroughly and keeping it for 24 hours can cure urea for its better efficiency.
- Fertilizer recommendations should preferably be based on crop sequence for multiple cropping. Application of FYM and phosphate should preferably be made in wet and dry season, respectively. Intercropping with green gram, blackgram, soybean, onion and groundnut prove highly remunerative in wide row crops like cotton and red gram as compared to pure crop stand and need no extra fertilizers.
- Balanced fertilization should be practiced based on the soil test.
- To the extent possible, using just enough water at different physiological growth stages as recommended for the crop being grown controls irrigation rate. It is better to give priority to irrigated fields as compared to unirrigated fields/crops as irrigated crops respond better to fertilizers.
- The introduction of leguminous crops in diverse rotational and inter-cropping sequence and use of bacterial and algal cultures play a very important role in meeting the fertilizer need of the crop.
- To the extent possible, green manuring with daincha (*Sesbania aculeata*) or sunhemp (*Crotalaria juncea*) should be practiced in low land paddy cultivation.
- For the compost made from straw and leaves having wide C:N ratio, add small quantity of N to increase N availability to young crop. Incorporation of finely ground rock phosphate or super phosphate with organic manures will make the manure more balanced.
- It is desirable to properly conserve and use organic waste. They should be incorporated 3-4 weeks before sowing the crop. It is preferable to apply FYM in kharif.
- Ensure proper plant spacing.
- Plant/sow the crop timely to get maximum benefit from fertilizers.
- Higher yielding varieties should be preferred over local varieties. Most responsive and best-suited crops should be selected to get maximum benefits of the limited quantity of fertilizers.

Chapter 12

Soil Fertility Evaluation

Methods of soil fertility evaluation

I. Nutrient deficiency symptoms on plants

II. Plant analysis

III. Biological tests

IV. Soil testing

V. Spectral sensing for site specific evaluation

Soil fertility evaluation

The diagnosis of the nutrient status of the soil by using different techniques or methods is known as soil fertility evaluation.

Methods of soil fertility evaluation

There are various diagnostic techniques that are commonly used to evaluate fertility of the soils. They are:

I. Nutrient deficiency symptoms of plants

It is a qualitative measurement of availability of plant nutrients. It is the visual method of evaluating soil fertility and diagnosing the malady affecting the plant. An abnormal appearance of the growing plant may be caused by a deficiency of one or more nutrient elements. The appearance of deficiency symptoms on plants has been commonly used as an index of soil fertility evaluation. If a plant is lacking in a particular element, more or less characteristics symptoms may appear. This visual method of soil fertility evaluation is very simple, not expensive and does not require elaborate equipments but it becomes difficult to judge the deficiency symptoms if many nutrients are involved. In such cases it requires experienced person to make proper judgment. The common deficiency symptoms are:

1. Complete crop failure at seedling stage.
2. Retarded/stunted growth.
3. Abnormal colour pattern. *e.g.*: Chlorosis (yellowing), necrosis (dying of tissue).
4. Malformation of different plant parts. *e.g.*: rosette appearance of leaves
5. Delayed maturity.
6. Poor quality of crops like low protein, oil, starch content, keeping/ storage quality reduced.
7. Internal abnormality like Hidden hunger (It is a situation in which a crop needs more of a given element, yet shows no deficiency symptoms).

II. Plant analysis

It is a valuable supplement to soil testing in the task of soil fertility evaluation. Plant analysis indicates the actual removal of nutrients from the soil and identifies nutrient status of plant and deficiency of nutrient element. It is a direct reflection of nutrient status of soil.

Advantages of plant analysis are;

a. Diagnosing or confirming the diagnosis of visible symptoms.
b. Identifying hidden hunger.
c. Locating areas of incipient (early stage) deficiencies.
d. Indicating whether the applied nutrients have entered the plant.
e. Indicating interactions or antagonisms among nutrient elements.

Plant analysis consists of three methods

Rapid tissue tests: It is a rapid test and qualitative or semi quantitative method. Fresh plant tissue or sap from ruptured cells is tested for unassimilated N, P, K and other nutrients. Examples of tissue tests- greenness index for chlorophyll for critical N concentration. ribonuclease activity for Zn and nitrogen reductase activity for nitrate.

The cell sap is added with certain reagents to develop color. Based on intensity of color low, medium and high color is categorized which indicates the deficiency, adequate and high nutrients in the plants respectively. It is mainly used for predicting deficiencies of nutrients and it is possible to forecast certain production problems.

1. Total analysis: It is a quantitative method and performed on whole plant or on plant parts. The dried plant material is digested with acid mixtures and tested for different nutrients quantitatively by different methods. The determination gives both assimilated and unassimilated nutrients such as nitrogen, phosphorus, potassium calcium, magnesium, suphur, iron, manganese, copper, boron, molybdenum, cobalt, chlorine, silicon, zinc, aluminum *etc*., in plants.

Most recently matured plant material is preferable for accurate analysis. Because critical nutrient concentration range (about 90% maximum yield) in plants at which crop yield is affected is relatively narrow, it is used for soil fertility evaluation.

Critical levels of nutrients in plants

Element	General range (%)	Critical level (%)
N	2.0 - 4.0	<2
P	0.2 - 0.5	<0.1
K	1.5 - 3.0	<1.0
Ca	0.5 - 3.0	<0.1
Mg	0.2 - 0.5	<0.2
S	0.2 - 0.5	<0.15
Fe	50 – 150 ppm	<5 ppm
Cu	5 – 20 ppm	<4 ppm
Zn	20 – 100 ppm	<15 ppm
Mn	20 – 500 ppm	<20 ppm
B	2 – 100 ppm	<20 ppm
Mo	0.1- 2.0 ppm	<0.1 ppm
Cl	0.2 - 2.0 ppm	-

III. Biological methods

It is conducted for calibrating the crop responses to added nutrients. Different methods are adopted for evaluating fertility status of soil.

1. **Field tests:** Field tests are conducted on different fertilizers and crops with treatments impositions in replications. The treatment which gives highest yield will be selected. These experiments are helpful for making general recommendations of fertilizer to each crop and soil and we can also choose right type and quantity of fertilizer for various crops. It is laborious, time consuming, expensive but the most reliable method. They are used in conjunction with laboratory and greenhouse studies as final proving technology and in the calibration of soil and plant tests. They are widely used by experiment stations.
2. **Indicator plants:** These are plants that are more susceptible to the deficiency of specific nutrients and develop clear deficiency symptoms if grown in that nutrient deficient soil. Hence these are called as indicator plants. Some indicator plants are;

Nutrients	Indicator plants
N, Ca	Cabbage, Cauliflower
P	Rape
K, Mg	Potato
Fe	Cauliflower, cabbage, potato, oats
Zn	Maize
Na, B	Sugar beet
Mn	Sugarbeet, oats, potato
Mo	Lucerne , Cauliflower
Cu	Wheat

3. **Microbiological test:** By using various cultures of microorganisms soil fertility can be evaluated. These methods are simple, rapid and need little space. Winogradsky was one of the first to observe that in the absence of mineral elements certain microorganisms exhibited a behaviour similar to that of higher plants. Microorganisms are sensitive to deficiency of nutrients and could be used to detect the deficiency of any nutrient. A soil is treated with suitable nutrient solutions and cultures of various microbial species (bacteria, fungi) and incubated for a few days. Then observing the growth and development of organisms in terms of weight or diameter of the mycelia pad, the amount of nutrient present in the soil is estimated.

Example:

a. *Azotobactor* method for Ca, P and K.
b. *Aspergillusniger* test for P and K
c. Mehlich's Cunninghamella (Fungus)- plaque method for phosphorus
d. Sackett and Stewart techniques (Azotobacter culture) to find out P and K status in the soil.
e. Laboratory and Green house Tests.

These are simple and more rapid biological techniques for soil fertility evaluation. Here, higher plants and small amounts of soils are used for testing. All these techniques are based on the uptake of nutrients by a large number of plants grown on a small amount of soil. It is used to assess availability of several nutrients and they are quantitatively determined by chemical analysis of the entire plant and soil. Some common methods are; a. Mitscherlich pot culture method for testing NPK status in oat, Jenny's pot culture test using lettuce crop with NPK nutrients, Neubauer seedling method for NPK and Sunflower pot-culture technique for boron.

IV. Soil testing

A soil test is a chemical method for estimating the nutrient supplying power of a soil. It is much more rapid and has the added advantage over other methods of soil fertility evaluation. One can determine the needs of the soil before the crop is planted. A soil test measures a part of the total nutrient supply in the soil.

Soil testing plays a key role in today's modern and intensive agriculture production system as it involves continuous use and misuse of soil without proper care and management. Soil analysis is helpful for better understanding of the soils to increase the crop production and obtaining sustainable yield. Soil testing is an indispensable tool in soil fertility management for sustained soil productivity.

Objective of soil testing

a. To evaluate fertility status of soil by measuring available nutrient status.

b. To prescribe or recommend soil amendments like lime and gypsum and fertilizers for each crop.

c. To assess nutrient deficiencies, imbalances or toxicities in soil and crop.

d. To test the suitability of soil for cultivation or gardening or orchard making

e. To know acidity, alkalinity and salinity problems.

f. To know morphology, genesis and classification of soil.

g. To find out the effect of irrigation on soil properties.

h. To prepare a soil fertility map of an area (village, Tehsil, district, state).

In the soil testing programme, "soil sampling" is the most important step to be followed for getting accurate results. Soil sampling is a process by which a true representative sample of an area or orcahrd can be obtained. The soil sampling must be done scientifically by adopting appropriate time and depth of sampling given for each crop for accurate analysis of soils.

Table: Interpretation of soil test results and critical levels of nutrients in soils

Nutrients	Low	Medium	High
Avail. Nitrogen (Kg ha^{-1})	280	280-580	>580
Avail.Phosphorus (P_2O_5 Kg ha^{-1})	22.5	22.5-55	>55
Avail. Potassium (Kg ha^{-1})	125	125-300	>300
Organic carbon (%) pH	0.5	0.5-0.75	>0.75
	<6.5 = Acidic		
	6.5-7.5 =Normal/Neutral		
	7.5-8.5 = Saline		
	>8.5 =Alkaline		
EC (dSm^{-1})	<0.80 = Normal		
	0.8-1.60 = Critical for some crops		
	1.6-2.5 = Critical for salt tolerant crops		
	>2.5 = Injurious to all crops		
Ca	<50% of CEC		
Mg	<4% of CEC		
S	<10 ppm		
Zn	<0.6 ppm (0.5-1.2 ppm)		
Fe	2.5-4.5 ppm		
Mn	<2 ppm		
Cu	<0.2 ppm		
B	<0.5 ppm (WS)		
Mo	<0.2 ppm		
Cl	<2 ppm (WS)		

v. Spectral sensing for site specific soil fertility evaluation

Remote sensing satellites are used to analyse the spectral signatures of crops and the fertility status is evaluated.

Chapter 13

Modern Approaches of Soil Fertility Evaluation and Fertilizer Recommendation

Soil fertility evaluation is conducted for providing an adequate supply of essential plant nutriens to ensure optimum plant productivity, while maximising economic benefits and minimising environmental degradation.

Mathematical modelling of nutrient application

The significance of an essential plant nutrient was contained in the Liebig's law of minimum.

Law

Plant growth would be limited by the nutrient, that is least available to a plant related to the plant's need (here clarified by the shortest stave) as long as it is no more limiting.

Fig. : Law of minimum. Here K is the limiting factor as its amount does not permit more growth.

When crop response to an essential nutrient present in soil or added to it increases in direct proportion, it can be expressed linearly as

$$Y = mx + C.$$

It applies to low soil fertility conditions and for nutrient uptake. In such cases the value of y- intercept 'C', is equivalent to the value of native availability of the nutrient 'm' is the slope of the curve or the unit increment and 'Y' is the yield or nutrient uptake and 'x' is the quantity of nutrient applied.

However, as the quantity of a nutrient increases, the unit increment in yield becomes smaller, and equivalently the crop yield reaches a maximum value. The crop response curve is described by Mitscherlich's equation or Mitscherlich-Spillman equation, $Y=A(1-e^{-c(x+b)})$ where 'x+b' is the total nutrient and 'b' is the plant available nutrient in soil, 'A' is the maximum yield attainable when x+ b is present in sufficient quantity and 'c' is the efficiency factor.

When the quantity of nutrient added to soil as well as that present in soil is in excess of crop requirements, the nutrient becomes toxic to plants or other factors become limiting, then the response curve obeys the polynomial equation like quadratic one

$$Y = ax+bx^2 + c$$

The maximum yield is given by the value of '- a/2b'. When Y is expressed as nutrient uptake, the value of 'c' estimates the quantity of available nutrient in soil.

Since the maximum yield varies with crops, seasons *etc.*, it is usually preferable to express plant productivity as relative of the maximum (% age).

With these fundamental concepts in mind two approaches have been developed-

1. Soil Test Crop Response (STCR)/Targeted Yield Concept
2. Diagnosis and Recommendation Integrated System (DRIS) Approach

1. Soil Test Crop Tesponse (STCR)

After introduction of high yielding varieties and hybrid crops, the need for systematic soil test crop response research in different soil agro-climatic regions become evident. ICAR established the AICRP on STCR in 1967 and the STCR concept was developed by Ramamoorthy Since 1987. STCR provides the relationship between a soil test value and crop yield. The soil test values are needed to be correlated with actual crop response obtained under field conditions. Separate calibration charts are needed for each crop and soil. Fertility gradient and regression approach and targeted yield concepts were evolved. This is also called as "rationalized fertilizer prescription approach" in which inherent soil fertility and yield level of the crop are taken in to account while recommending the fertilizer doses.

Objectives of STCR

1. To prescribe fertilizer doses for a given crop based on soil test values to achieve the "Targeted yields" in a specific soil agro-climatic region under irrigation or protective irrigation conditions by using mathematical equations for different crops and different soil agro-climatic zones separately.
2. This takes in to consideration-the efficiency of utilization of soil and added fertilizer nutrient by the crops and its nutrient requirements for a "desired yield level".

Concept of STCR

STCR approach aims at obtaining a basis for precise quantitative adjustment of fertilizer doses under varying soil test values and response conditions of the farmers and for targeted levels of crop production. These are tested in follow up verification by field trials to back up soil testing laboratories for their advisory purpose under specific soil, crop, and agro climatic conditions. The fertilizers are recommended based on the following criteria.

- Fertilizer recommendations based on regression analysis approach
- Recommendations for certain % of maximum yield

STCR methodology takes in to account the three factors;

i) Nutrient requirement (NR) in kg/ quintal of the produce,

ii) Percentage contribution from soil available nutrients (SE) and

iii) Percentage contribution from added fertilizers towards making effective fertilizer prescriptions for specific yields.

i) Nutrient Requirement (NR) = Total nutrient uptake (grain+straw) kg ha (in kg for producing 1 Quintal grain) Yield of grain (q/ha)

$$\text{ii) Fertilizer Efficiency (FE in \%)} = \frac{\text{Total nutrient uptake (kg/ha)} - \text{soil test value in treated Plot (kg/ha)} \times \text{soil nutrient efficiency (\%)}}{\text{Fertilizer dose applied (kg/ha)}} \times 100$$

$$\text{iii) Soil Efficiency (SE in \%)} = \frac{\text{Total nutrient uptake in control kg/ha}}{\text{Soil test value in control kg/ha}}$$

1) With the help of above parameters, adjustment equations have been developed for a number of crops in various soils.

E.g.: For potato at Bhubaneswar

General fertilizer recommendation : 120-60-120.

a. Fertilizer N = 1.8 T – 1.1 Soil N
b. Fertilizer P_2O_5= 0.5 T – 1.8 Soil P
c. Fertilizer K_2O = 1.1 T – 1.3 Soil K_2O

Where,

T= Targeted yield of potato

Advantages

1) Efficient and profitable site specific fertilizer recommendation for increased crop production and for maintenance of soil fertility.
2) Provides balanced, efficient and profitable nutrient application rates for pre-set yield targets giving due consideration to basic fertility status of soil.

Targeted yield concepts

These are soil test based recommendations but given for different yield goals and not for a single optimum yield level. A large variety of fertilizer prescription have been made available by putting soil test values in to certain mathematical equations and finding out the amounts of nutrients needed for a given yield target.

2. Diagnosis and Recommendation Integrated System (DRIS) Method

Concept of DRIS

DRIS is a new approach to interpreting leaf or plant analysis which was first developed by "Beaufils" (1973) named as Diagnosis and Recommendation Integrated System (DRIS). It is a comprehensive system which identifies all the nutritional factors limiting crop production and increases the chances of obtaining high crop yields by improving fertilizer recommendations. The DRIS method uses "nutrient ratios" instead of absolute and or individual nutrient concentrations for interpretation of tissue analysis. There is a set of optimum ratios among the nutrient elements (N/P or N/K or K/P or N/S, N/P/K/S) within a given plant that remains uniform throughout the growth period. DRIS mainly uses the "nutritional balancing" concept (Relationship among nutrients) in the detection of nutritional deficiencies or excess in the plant. Nutrient balance is a part of the proper interpretation of DRIS system because nutrient interactions to a larger extent determine crop yield and quality. The nutrient ratios are helpful to obtain special indices which are called "Nutrient Index" or "Beaufills Nutrient Indexes" (BNI). The nutrient index values are used to rate the nutrients in order of their need by the plants analyzed. It also measures how far particular nutrients in the leaf or plant are from optimim are used in the calibration to classify yield factors in order of limiting importance. BNIs are actually expression of the supplies of nutrients relative to each other. The concentration of each nutrient in the plant has an effect on the index value for each of the other nutrients. An abnormally high concentration of one or more nutrient (s) will decrease the index values of other nutrients. There will be positive and negative values for the nutrient index. The nutrients with positive indexes appeared to be in "excess" and nutrients with negative indexes appeared to be "deficient" in plants. DRIS indices can be calculated individually for each nutrient using the average nutrient ratio and the deviation obtained from the comparison with the optimum value of a given nutrient ratio. DRIS is a mathematical technique to apply plant analysis information (Nutrient concentration) for diagnosing the most limiting nutrient in a production system. The evaluation is made by comparing the relative balance of nutrient content with norms established for that crop under high yield conditions. The evaluation is made by comparing the relative balance of nutrient content with norms established for that crop under high yield conditions. To develop a DRIS for a crop, the following requirements must be met whenever possible:

1. All factors suspected of having effect on crop yield must be defined
2. The relationship between these factors and yield must be described
3. Calibrated norms must be established
4. Recommendations suited to particular sets of conditions and based on correct and judicious use of these norms must be continually refined.

Advantages

- The importance of nutritional balance is taken in to account in deriving the norms and making diagnosis. It helps quantify the nutrient balance in the plant.
- The norms for nutrient content in leaves can be universally applied to the particular crop.
- Diagnosis can be made over a wide range of stages of crop development.
- The nutrient limiting yield through either excess or insufficient can be readily identified and arranged in order of their limiting importance for yield.

The balance of nutrients within a plant is often more important than individual nutrient concentration. The DRIS approach focuses on nutrient balance as an alternative to critical concentration approach to plant analysis interpretation. The ratio of mobile nutrients like N,P, K,S is mostly constant throughout the growing period but there is no universal benchmark.

Chapter 14

Fertilizers

Definition of Fertilizer

Classification of Inorganic Fertilizers

I. Nitrogenous fertilizers
II. Phosphatic fertilizers
III. Potassic fertilizers
} Primary nutrient fertilizers

IV. Secondary nutrient fertilizers

V. Micronutrient fertilizers

Fertilizer

Fertilizers are defined as materials having definite chemical composition with a high analytical value that supply essential plant nutrients in available form. They are usually manufactured by industries and sold with a trade name. They are commonly synthetic in nature and also called as chemical fertilizers/inorganic fertilizers/commercial fertilizers other than lime and gypsum. Most of the chemical fertilizers are inorganic in nature. The only exception to this is urea and calcium cyanamide ($CaCN_2$), the solid organic N fertilizer. In India the use of artificial fertilizers was first initiated in 1896 when imported Chilean nitrate was used as a fertilizer. Presently fertilizers have become an integral part of agricultural economy as they increase the fertility of soils and enable them to support high yields. About 50% of the increase in crop production during recent times has been attributed to fertilizer use; though the fertilizer use efficiency is very poor.

II Classification of Fertilizers

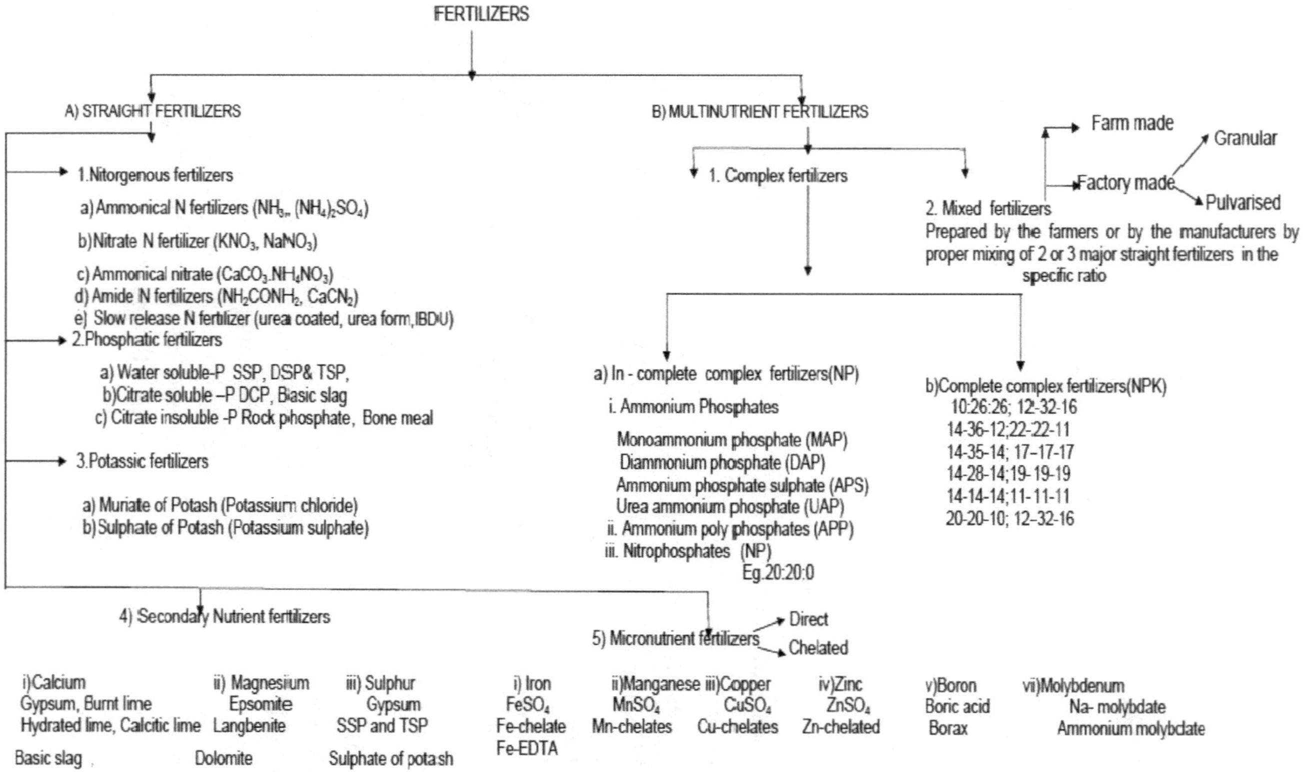

Classification of inorganic fertilizers

A. Based on number of nutrients present

1. **Straight fertilizers:** Are those fertilizers containing or supplying only one plant nutrient element at a time. For *eg*., Urea, Ammonium Sulphate (NH_4SO_4), Ammonium nitrate (NH_4NO_3), Single super phosphate (SSP), Muriate of potash (MOP-KCl).
2. **Complex fertilizers:** Fertilizers containing at least 2 or more of the primary essential nutrients (NPK). They are chemical mixtures, granular and free flowing and easy to apply.

There are two types of complex fertilizers;

a. **Complete or compound fertilizer**: They are the chemical mixtures of three or more primary or major nutrient elements (NPK) in one compound or mixture. They are usually in granular form and easy to apply. *Ex:* 10-26-26, 17:17:17, 19:19:19.

b. **Incomplete complex fertilizers**: A fertilizer material lacking any one of three major nutrients or containing only two of the primary nutrients like N, P and K.

Ex: N-P complex fertilizer:-Nitro-phosphate (Suphala:-15-15-0, 20:20:0)

-Diammonium phosphate: 18-46-0

Characteristics of complex fertilizers (CF)

1. They usually have a high content of plant nutrients. As such they are also called high analysis fertilizers.
2. They usually have a uniform grain size, granular form and good physical condition during storage.

Advantage of complex fertilizers

1. In one application we can supply more nutrients and need not apply separately.
2. Balanced nutrition can be achieved.
3. Less cost is involved in transportation and application.
4. They are available in different grades according to need of the soils and crops.
5. Being granular, it is easy to apply by broadcasting.
6. Some complex fertilizers also provides some micronutrients to soil.
7. Transport and distribution is easy.
8. They are non-caking and non- hygroscopic, thus safer for storage

3. Mixed Fertilisers or Fertiliser Mixtures

A mechanical/physical mixture of two or more straight fertilizer materials in suitable proportion is referred to as fertilizer mixture or mixed fertilizers. Sometimes, complex fertilizers containing two plant nutrients are also used in formulating fertilizer mixtures. Specific fertilizer grades are recommended for specific crops depending upon the soil and climatic conditions of the region. The mixed fertilizers are usually in powder form or sometimes granular form. Fertilizer mixture (FM) are free flowing and easy to apply. The mixed fertilizers can be made according to the need of the crop and there is wide scope for adjusting the fertilizer ratio..

Guidelines for mixing fertilizers

Some fertilizers cannot be mixed with other fertilizers. Mixing of incompatible fertilizer leads to a loss of some of the nutrients in the form of gas, converting soluble nutrients into insoluble form or caking. Certain fundamental principles are to be followed in mixing fertilizers are.

1. Ammonium sulphate, ammonium chloride and other ammonical fertilizers and nitrogenous organic manures should not be mixed with lime.
2. Urea should not be mixed with Super phosphate (SP)
3. Calcium cyanamide, basic slag, quick lime slaked lime should not be mixed with N in NH_4N form.
4. Super phosphate should not be mixed with lime or $CaCO_3$ or wood ashes.
5. $NaNO_3$ or KNO_3 should not be mixed with Super phosphate.
6. Ammonium sulphate, nitrate should not be mixed with lime.
7. Nitrochalk should not be mixed with SP or lime.

The commonest fertilizer mixture can be made from SSP, Ammonium sulphate, SOP, Bonemeal, and MOP.

Advantages of fertilizer mixtures

1. Less labour is required to apply fertilizer mixture to soil. Individual crop wise fertilizer mixture can be made.
2. Balanced nutrition can be achieved.
3. The residual acidity of fertilizers can be effectively controlled by adding liming materials in the mixtures.
4. Micronutrients can be incorporated in fertilizer mixtures.
5. They have a better physical condition and more easily applied.
6. There is no need of purchasing straight fertilizers separately.

Disadvantages of fertilizer mixtures

1. Does not permit application of individual nutrients according to the needs of crops during specific times.
2. The unit cost of plant nutrients is higher than of straight fertilizer.
3. Lack of knowledge about proper mixing and their use.
4. Fertilizer mixture of particular grade suitable for particular crop cannot be applied for all crops.

Classification of fertilizers based on particular plant nutrient element

The element forms their principal constituent in the fertilizer.

1. Nitrogenous fertilizers } Primary nutrient fertilizers
2. Phosphatic fertilizers }
3. Potassic fertilizers
4. Secondary nutrient fertilizers
5. Micronutrient fertilizers.

I. **Nitrogenous fertilizers**: There are 6 groups.

a. Ammonium fertilizers

i. Ammonium sulphate $(NH_4)_2\ SO_4$– 20% N
ii. Ammonium chloride : NH_4Cl_2 -24-26%
iii. Ammonium phosphate : $NH_4H_2PO_4$ -20%N + 20% P_2O_5 or 16% N and 20% P_2O_5
iv. Anhydrous ammonium (82%N)
v. Ammonium solution- 20-25%N
vi. Ammonium carbonate- 21-24%N
vii. Ammonium bicarbonate- 17%N

b. Nitrate fertilizers

i. Sodium nitrate or Chilean nitrate : $NaNO_3$– 16%N
ii. Calcium nitrate: $CaNO_3$– 15.5% N
iii. Nitrophosphate

c. Both Ammonium and nitrate fertilizers

i. Ammonium nitrate: NH_4NO_3- 33-34%N
ii. Calcium ammonium nitrate (CAN) -25, 26 and 28% N
iii. Ammonium sulphate nitrate (ASN) - 26%N

d. Amide fertilizers

i. Urea – 46% N
ii. Calcium cyanamide- 21 %N
iii. Urea phosphate
iv. Urea sulphate

e. Nitrogen solutions

i. Anhydrous ammonia
ii. Aqueous ammonia
iii. Solution containing one or more of the following: urea, ammonium nitrate, ammonia

f. Slowly available nitrogenous fertilizers

i. Urea formaldehyde compounds
ii. Neem Cake coated Urea (NCU)
iii. Lac Coated Urea (LCU)
iv. Sulphur Coated Urea (SCU)
v. Urea super granules (USG)
vi. Prilled urea (PU)

II. Phosphatic fertilizers

They are broadly classified into 3 major groups on the basis of their solubility either in water or in citrate or citric acid.

a. **Water soluble phosphatic fertilizers**(Contain phosphoric acid or mono calcium phosphate). *Ex*;

1. Single super phosphate (SSP) - 16-18 % P_2O_5
2. Triple super phosphate (TSP) - 46-48 % P_2O_5
3. Double super phosphate (DSP) - 32% P_2O_5
4. Di-Ammonium phosphate (DAP) - 18%N and 46% P containing dicalcium phosphate

b. Citric acid soluble phosphatic fertilizers

1. Dicalcium phosphate (DCP) - 34-39% P_2O_5
2. Rhenamia phosphate- 23-26% P_2O_5
3. Basic slag- 14-18% P_2O_5
4. Raw or steamed bone meal- part of P_2O_5 soluble in citric acid
5. Fused calcium magnesium phosphate- 16.5% P_2O_5.

c. Water insoluble or citric acid insoluble phosphatic fertilizers. Containing tricalcium phosphate ($Ca_3(PO_4)_2$

Ex: Rock phosphate - 20-40% P_2O_5,

Raw bone meal - 20-25% P_2O_5

Steamed bone meal - 22% P_2O_5

Pyrophos - 17% P_2O_5.

III. Potassic fertilizers

A. Fertilizers having K in the Chloride form

1. Muriate of Potash (MOP) – KCl- 60-62 % K_2O

B. Fertilizers having K in Non-chloride form

1. Sulphate of potash (SOP) -K_2SO_4: 48-52% K_2O
2. Potassium nitrate KNO_3 - 44% K_2O

IV. Secondary nutrient fertilizers

Secondary elements are as important as primary elements because they help in uptake of primary elements by plants. They are required in very little quantity as compared to primary elements. The most important secondary nutrients are Ca, Mg and S. The fertilizers carrying secondary nutrients are;

1. Calcium cyanamide (39.57% Ca)
2. Calcium ammonium nitrate (8.0% Ca& 4.5% Mg)
3. Calcium nitrate (1.5% Mg)
4. Super phosphate (20.0% Ca)
5. Bone meal (23.1% Ca)
6 Limestone (32.58% Ca)
7 Dolomite (20.0% Mg)
8 Gypsum (29.40% Ca& 21.0% S)
9. Potassium sulphate (18.5% S & 0.6 to 0.9% Mg)
10. Ammonium sulphate (24% S)

V. Micronutrient fertilizers

Micronutrients are those which required by plants in very minute quantities by plants but they have equal role as that of primary nutrients. They govern most of the physiological as well as biochemical reactions of plant growth and development. The most important micronutrients are iron, manganese, zinc, copper, molybdenum, chlorine, boron and nickel. The fertilizers carrying micronutrients are;

Fe carrying fertilizers

1. Ferrous sulphate (19.0% Fe)
2. Ferric sulpahte (23.0% Fe)
3. Ferrous ammonium sulphate (29.0% Fe)
4. Ferric and ferrous oxide (70.0 and 77.0% Fe)

Boron carrying fertilizers

1. Borax (11.0% B)
2. Boric acid (17.0% B)
3. Sodium tetra borate (14.0% B)
4. Borosite (21.0% B)

Manganese carrying fertilizers

1. Manganese sulphate (20.0 to 28.0% Mn)
2. Manganese carbonate (31.0% Mn)
3. Manganese chloride (17.0% Mn)

Zinc carrying fertilizers

1. Zinc sulphate (55.0% Zn)
2. Zinc oxide (67.0% Zn)
3. Zinc sulphide (67.0% Zn)
4. Zinc ammonium sulphate (33.5% Zn)

Molybdenum carrying fertilizers

1. Sodium molybdate (39.0% Mo)
2. Ammonium molybdate (54.0% Mo)

Copper carrying fertilizers: Copper sulphate.

Chapter 15

Methods of Fertilizer Application

Methods of fertilizer application

Two major methods are used to apply fertilizers

I) Application of fertilizers in solid form

II) Application of fertilizers in liquid forms

Basic principles of fertilizer application

Fertilizers are chemically reactive substances and costly inputs which need to be applied carefully either to soil or foliar by adopting scientific methods and principles. Various methods have been used to apply fertilizers to soil or crops and their advantage along with disadvantage and principles is to be understood.

Application of fertilizers in solid form

1) Broadcast

a) Broadcasting at sowing/planting or before sowing/ planting time

b) Broadcasting at later stages of crop growth *i.e.*, top dressing

2) Placement

a) Plough sole placement

b) Deep placement

c) Sub soil placement

3) Localized placement

a) Contact or Combined drilling or drill placement

b) Band or hill placement

i) Spot placement,

ii) Row placement,

iii) Side dressing,

iv) Pellet application

c) Basin application

Application of fertilizers in liquid forms

a) **Starter solution:** mixed with ratio of 1:2:1 or 1:1:2 of NPK respectively for vegetable. Seedlings at the time of transplanting.

b) **Foliar application or spray fertilization:** Spraying the nutrient solution with appropriate concentration over the foliage.

c) Direct application to the soil or plant (injection to soil or plant).

d) **Fertigation:** Application fertilizer through irrigation water (Sprinkler or drip method).

I) Application of fertilizers in solid form

Solid fertilizers of straight or complex or mixed NPK fertilizers are directly applied to soil by adopting different methods of application based on the type of crop, season, age of the crop, type of fertilizer and their concentrations.

1. **Broadcasting:** It is a process where the fertilizer is spread over the entire soil area evenly and uniformly. This may be done before the land is ploughed, before planting or while the crop standing. It may be of two types:
   a. **Broadcasting at planting or sowing:** During sowing time, fertilizers are uniformly spread over the soil and mixed properly.
   b. **Broadcasting as top dressing:** Followed for closely spaced crops. It may be made at the time of critical crop growth period. It is essential to note that the fertilizers are not to be applied when the leaves of plants are wet. This may create injury to the leaves or direct fall of fertilizer granules like urea or KCl on leaves leads to leaf scorching.

Advantage

Easy, simple, less cost (Labour), no special equipment, large quantity of fertilizer can be applied.

Disadvantages: (Over placement)

1) Uneven distribution of fertilizer which causes a ranked (un uniform) growth of crop.
2) Loss of nitrogen through volatilization, erosion.
3) It requires relatively large quantity of fertilizers.
4) Costly fertilizers cannot be applied by this method, requires experience and skill of application.

5) Fixation of nutrients especially PO_4 and K due to contact of large volume of soil compare to that of placement.
6) Stimulates the more weed growth.

2. **Placement**: Fertilizers are placed in the soils irrespective of the position of seed, seedling or growing plants before sowing or after sowing the crops. The following methods are most common;
   a) *Plough*: sole placement: It is applied in a continuous band at the bottom of the furrow during the process of ploughing.
   b) *Deep placement*: It is specially followed in paddy fields for placement of N fertilizers

      ($(NH_4)_2 SO_4$ and urea) in the reduced zone. It prevents loss of ammonia and makes the nutrient easily available to crop. The fertilizer is applied under the plough furrow in the dry soil before flooding the land and making ready for transplanting.
   c) *Sub soil placement*: It is recommended in humid and sub humid regions where sub soils are strongly acid. P & K fertilizers are used.
   d) *Localized placement*: Application of fertilizers in to the soil close to the seed or plants in bands or in pockets. Only small quantities of fertilizers are applied. It reduces fixation of P & K. It includes-
      i) Drill placement or contact placement or combined drilling: Drilling of seeds and fertilizers together while sowing in the same row. Old method and useful for PO_4 and potash fertilizers in rainfed areas.
      ii) *Band or hill placement*: When the plants are placed 3 feet or more on both sides fertilizers are placed close to plants in bands on one or both sides of the plant. It is practiced for N &P Fertilizers to orange, banana, papaya, apple, pear, coconut, cashew nut and other fruit trees.
      iii) *Row placement:* Fertilizers are applied in rows
      iv) *Side dressing:* Fertilizers are applied between the rows (followed for Vegetables and fruit crops).
      v) *Pellet application:* Make pellet form (1:10 ratio of fertilizer (N) & soil and make it to dough) and apply to soil.

Advantages of placement/banding

1) Fixation of nutrients appreciably reduced.
2) Never harms plants and seeds.
3) No weed growth (weeds do not get nutrients).
4) There is no volatilization loss of N.

5) It gives better residual effects and every unit of nutrient is effectively utilized (better FUE).
6) Smaller quantities of fertilizers can be effectively applied; both solid and liquid fertilizers are applied.
7) It is best method for P & K application.
8) It is useful for widely spaced crops, quick growing crops, dry land crops.

Disadvantage

High labour cost and requires specialized equipments.

II. Application of fertilizers in liquid form

Solid form of fertilizers with known concentrations are dissolved in definite volume of water and sprayed over the leaves. Commercially liquid fertilizer solutions are available in the market may be applied to the plants or soil directly. Various methods are used to apply the fertilizer in liquid form.

i) *Starter solution*: Fertilizer solutions are prepared by dissolving the fertilizers in water at the ratio of 1:2:1 or 1:1:2 of N: P_2O_5: K_2O and applied to young vegetable plants at the time of transplanting.

   It is used in place of watering which helps in better establishment & early growth of plants. Only small amount of fertilizer is needed.

ii) *Foliar application*: Spraying over the leaves of growing plants with suitable fertilizer solutions of lower concentrations (recommended) to supply one or combination of nutrients. Most of the micronutrient fertilizers (Zn or Boron) are applied as foliar spray and urea is also used for foliar spray.

iii) *Direct application of Liquid fertilizers to soils*: Liquid anhydrous ammonia or N, P, K fertilizers are applied to soils (10 cm depth) with the help of special equipment. Plant injury or wastage of ammonia is very much limited.

iv) *Fertigation*: It is the process of direct application of liquid fertilizers through irrigation water. Straight and mixed fertilizers of N P K which are easily soluble in irrigation water can applied through furrow or drip irrigation or sprinkler irrigation system.

Advantages of liquid forms of fertilizers

1) Under acute deficiency of elements like N, P, K, and micronutrients during active growth period of plants can be corrected.
2) Rapid plant absorption, quick response and for fruit trees more efficient and effective method.

3) Along with fertilizer solutions we can apply insecticides and herbicides with sticking agents and also some soil amendments like lime and gypsum.
4) Small amount/concentration of nutrients can be used.

Disadvantage of liquid forms

1) Sometimes fertilizer solutions causes scorching or burning of leaves due to presence of acids in it like ammonium sulphate or chlorides in KCl fertilizer.
2) It is more expensive/costly.
3) Several applications are needed for moderate to high fertilizer rates.
4) All the fertilizers cannot be applied. *Ex*: Ammonium sulphate is unsuitable for foliar spray.

Basic principles of fertilizer application

1) Nitrogen fertilizers must be applied in split doses so that the N loses through leaching and washing could be reduced as nitrogen being readily soluble and highly mobile.
2) Except in acidic and highly alkaline soils, the PO_4 must be applied in one dose as basal placement but in acidic soils RP, bone meal or basic slag may be applied at least a fortnight before sowing or crop planting whereas in alkaline soils spraying of PO_4 has given better results.
3) The potassic fertilizers should be applied in a single dose but split application along with N as top dressing has given better response in heavy soil types.
4) Sandy soils need split application of N and K.
5) A combination of organic manures and fertilizers is always beneficial for achieving highest nutrient recovery and best fertilizer use efficiency.
6) Mixed or complex fertilizers should be applied as basal before sowing or planting.
7) Soil amendments and soil conditioners should always be applied at least one month before sowing or planting of crops through broadcasting or proper mixing in to the soil.

Chapter 16

Fertilizer (Control) Order, 1985

Amendment to the Fertiliser (Control) Order, 1985 w.r.t. Bulk Sale of City Compost Under

D.O No. MD-SBM/FG/10/2017 18 February 2017

Fertilizer (Control) Order, 1985 which is administered by Deptt.of Agriculture Cooperation, Govt. of India has been issued under the Essential Commodities Act, 1955. The FCO lays, down as to what substances qualify for use as fertilizers in the soil, product-wise specifications, methods for sampling and analysis of fertilizers, procedure for obtaining license/registration as manufacture/ dealer in fertilizers and conditions to be fulfilled for trading thereof, *etc.*

Quality control of fertilizers

Objectives: Laws and regulations governing the manufacture and sale of fertilizers are imperative in order to check

1. Spurious standards and adultrated fertilizers entering in to the market.
2. To ensure quantity of nutrients and quality of carriers present in the fertilizer.
3. To ensure quantity of nutrients and quality of carriers present in the fertilizer.
4. To eliminate black marketers off the market.
5. The total N,P_2O_5 and K_2O must be guaranteed in terms of percentage of each of these nutrients in a given fertilizer. The oxide expression for P and K are actually in accurate and confusing as they are based on early practices with which chemists determined the elements by ignition and weighing the oxides. The current practice is to express the elemental concentration of these nutrients.

By considering the above said objectives, the government of India passed The Fertilizer Control Order [FCO] on 28th of April, 1957 in exercise of the power conferred by the section III of the essential commodities act of 1955. This order is intended to regulate the manufacture, distribution and supply of the fertilizers in India at a control cost. This has been effective from May 18th 1957. It is revised in 1985 with effect from 25-9-1985. The Government of India has delegated to powers to State Governments to implement the order.

The Government of India (G O I) also passed the Fertilizer. Movement Order (FMO) on 31st December, 1960 in order to regulate the inter state movement of fertilizers and the export of fertilizers which came into force with effect from 1-1-1961.

Fertilizer control order (FCO) Regulations

1. All the fertilizer manufacturer should obtain licence from the Commissioner of Agriculture, State Government concerned for the manufacture of fertilizer and mixed fertilizers.
2. The fertilizer dealers should on renewable basis, register their dealership with the Assistant Director of the Agriculture (ADA) Regular of the division concerned in a state.
3. The terms and conditions of manufacture, distribution and sales imposed by the government should be followed.
4. Duties of inspecting officers and the dealers are specified.
5. Fertilizer specifications and kind of package are stated.
6. Method of drawing fertilizer samples for analysis in the fertilizer testing laboratories is stated.
7. Powers are vested with the FCO enforcing officials to book the cases against the fraudulent manufacturers, distributors and dealers of fertilizers.

Specifications and standards for important fertilizers [As per FCO,1957]

1 **Urea** Moisture per cent by weight 1.0 maximum

Total nitrogen per cent by weight 44.0 minimum

Biuret per cent by weight 1.50 maximum

Particle size :In the form of granule the material shall pass through 2.8 mm and not less than 80% by weight shall be retained on 1mm.

2 **SSP** 1. Moisture per cent by weight, maximum 12

2. Free phosphates as (P_2O_5) per cent by weight 4.0

3. Water soluble phosphates (as P_2O_5) by weight maximum 16.0.

3 **MOP** Moisture per cent by weight maximum 12.00.

Water soluble potassium per cent by weight minimum 60.

4 **DAP** Moisture per cent by weight 1.0

Total nitrogen per cent by weight minimum 18

Total phosphates (as P_2O_5) per cent by weight minimum 46.0

Water soluble phosphates (as P_2O_5) per cent by weight 41.6.

D.O.No. MD-SBM/FG/10/2017 18 February 2017

Amendment to the Fertiliser (Control) Order, 1985 w.r.t. Bulk Sale of City Compost

Dear Sir

As you are aware, the Government of India is implementing 'Swachh Bharat Mission' (SBM) since October, 2014 with the goal of achieving a Clean India by 2nd October 2019. A key component of the mission is to help cities across to manage Municipal Solid Waste (MSW) in a sustainable manner through promoting efficient mechanisms for waste segregation, collection, storage, processing and final disposal.

2. In its meeting on 23rd January 2016, the Union Cabinet has approved the policy for promotion of city compost, wherein a Market Development Assistance (MDA) of Rs. 1500 per metric tonne of city compost will be paid to the fertilizer marketing companies, which are obligated to purchase all the compost manufactured by respective cities to which they have been tagged.
3. Following the release and implementation of this policy, it was identified that ULBs apart from manufacturing compost would like to market it on their own. Therefore, to provide further impetus to the existing policy and to increase the volume of sale, an amendment was released on 28.09.2016, wherein **ULBs can now also market compost directly to farmers in bulk and claim market development assistance of Rs.1500 per tonne.** It is anticipated that this policy will reduce the cost incurred by manufacturers in bagging and transportation of compost reducing the price at which compost will be made available to farmers, thereby boosting the sales of compost.
4. As per Clause 22 c. of THE FERTILISER (CONTROL) ORDER 1985 wrt. Bulk sale of fertilisers: "the Central Government may by notification published in the Official Gazette in this behalf authorise a manufacturer/ importer to sell any fertiliser manufactured/ imported by him In bulk also direct to farmers for such period as may be specified in that notification: Provided that a certificate indicating the minimum guaranteed percentage

of plant nutrients is issued by the manufacturer/importer to each farmer at the time of such sale"

5. The proposal for amendment of this clause, for **exempting the Municipality from the requirement of notification for bulk sale of fertilizer (in this case compost) has been proposed and approved by a competent authority** vide DO F. No. 8-2/2008 Fert Law (See Enclosed).
6. The management of solid waste through collection, processing, transportation and disposal in India is the responsibility of urban local bodies (ULBs) which is carried out either on their own or in partnership with other entities.
7. Given the above, **all entities including ULB run compost plants are exempted from the requirement of notification for bulk sale of compost as mentioned in the circular above. These plants can register on the mFMS portal** (http://mfms.nic.in/) **to log in production details and claim the subsidy as mentioned above.**

Praveen Prakash

Chapter 17

Biofertilizer

Use of biofertilizers in agriculture and their advantage

Bio-fertilizers or bio-inoculants

These are productions containing living organisms such as nitrogen fixers or phosphorus solubilises which are useful for agricultural production. Biofertilizers play a significant role in improving soil fertility and plant growth. By fixing atmospheric nitrogen both in association with plants and without plants solubilize insoluble soil phosphates into soluble phosphates there by increasing the availability of phosphorus secrete growth promoting substances and supplies to the inoculated plants.

Verma & Bhattacharyya (1994) classified the bio fertilizers based on the nutrient provided, which is as follows:

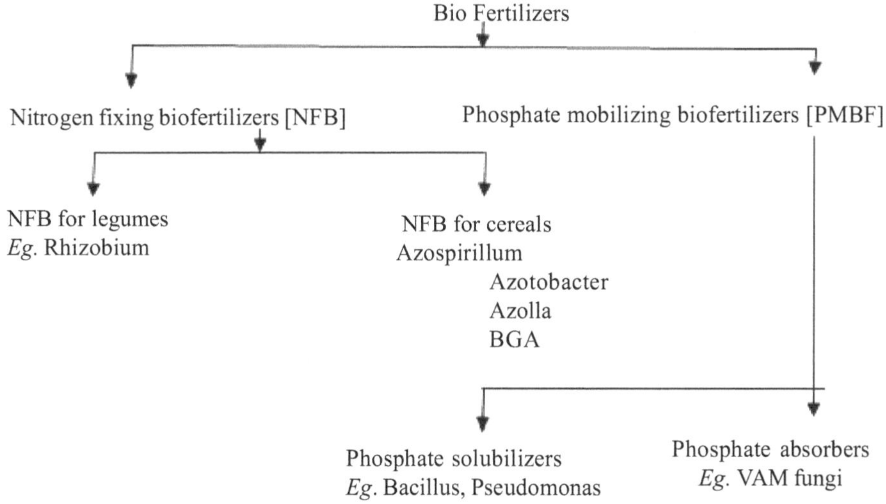

I. *Rhizobium* inoculant

*Rhizobium b*elongs to the family Rhizobiaceae and symbiotic in nature. It establishes symbiotic relations with pulses and other legumes and can fix 50 to 100 kg N ha^{-1} by which application of chemical nitrogen fertilizers could be reduced and supplemented nearly 80-90 per cent of nitrogen requirement of crop is met by nitrogen fixation in root nodules of legume crops. It is used for pulse legume crops like red gram, chickpea, lentil, black gram *etc*., and oil seed legumes like soybean and groundnut and forage legumes like besrseem , lucerne and pillipesara. On average yield increase was 15-30 % due to *rhizobium* inoculation.

The rhizobia show host specificity to some extent. In other words rhizobia isolated from groundnut can not be used for red gram. Particular rhizobium species has the ability to form nodules only on roots of limited and related plant species. Again effective strains are to be selected for effective nodulation. The response of inoculation is governed by

1. Host plant 2. Strains effectiveness 3. Soil type and 4. Rain fall

Method of preparation

This biofertilizer is available commercially as carrier based (lignite or peat) biofertilizers. Direct application of rhizobium to seed is the most common form of legume inoculation (seed coat). However, for groundnut, slurry of peat based inoculam in the seed furrows is recommended as the seeds are fragile. The recommended quantity of rhizobium fertilizer is 2-3 packets/ha (packet: 200 grams).

II. *Azotobacter* inoculant

Azotobacter belongs to the family Azotobacteriaceae, which is a chemoautotrophic in nature, free living, non-symbiotic in nature and is recommended for application to non-legume crops. It thrives well in many types of soils, with a minimal dose of phosphorus and carbon compounds. It grows best in neutral to alkaline soils. By the application of Azotobacter to various crops, the amount of recommended doses of nitrogenous fertilizers can be reduced by 10-20%. It produces plant growth promoting substances like vitamins of B group, Indole acetic acid, Gibberellic acid by which plants grow quickly and healthy. Yield increase of 10-15% was noticed in some millets. This biofertilizer is recommended for rice, wheat , millets and other cereals, vegetables, sunflower, mustard, tobacco sugarcane, cotton etc. and also fruit crops.

Method of usage: It is carrier based biofertilizer (lignite / peat). *Azotobacter* can be applied by seed inoculation, seedling inoculation, or by broad casting in the field after mixing with FYM or with well decomposed compost.

Seed inoculation: Seed is treated with Jaggery to make it sticky and then treated with biofertilizer and shade dried.

Seedling inoculation: Make slurry of biofertilizer and dip the roots in it. The recommended dose of Azotobacter is 2-3 packets/ha.

III. *Azospirillum* inoculant

It belongs to the family spirillaceae, chemoautotrophic and associative in nature. It is an associative nitrogen fixing organism was found to be beneficial for sorghum, wheat, maize, barley, fodder grasses and minor millets. Yield increase was 15-30 %. By the application of this fertilizer nitrogen saving is 20-30 kg/ha. It also produces growth regulating substances.

Application of this biofertilzer results in

1. Increased mineral uptake
2. Increased water uptake
3. Root development
4. Vegetative growth
5. Make the plant drought resistant

Note: Principle and mode of application is that of Azotobacter

IV. Azolla [Botanical name is *Azolla pinnata*]

1. It belongs to the family Azollaceae. *Azolla* is a water fern. It is considered as aquatic weed commonly found in floating in idle ponds, tanks shallow ditches and canals. Azolla is associated with rice field. A blue green algae (BGA) *Anabaena azollae* living in the epidermal cavity of lower side of the leaf of Azolla. The symbiotic association of *Azolla pinnata* and *Anabaena azollae* is termed as Azolla Anabaena complex. This algae fixes atmospheric nitrogen for Azolla and in exchange, the plant provides shelter and food to the algae. As this complex fixes atmospheric nitrogen, it has great potentiality for use, in agricultural field as biofertilizer and can be used as an alternative to nitrogen fertilizers. It is recommended as green manure because of its large biomass and high N content (46%N) for sub merged rice fields, within maximum temperature of 38°C. The yield increase is 15-20%.

Ideal conditions for growth and multiplication of *Azolla*

2. Depth of water should be maintained at 8-10 cm in the field.
3. Temperature in between 14 to 35 °C. Optimum for quick growth and multiplication is 20 to 28 °C.

4. Suitable soil pH is 6.0 to 8.0.
5. Single Super Phosphate should be applied at the rate of 20-25 kg ha^{-1}.
6. Furadon granules @ 2-3 kg/week should be applied to control leaf eating insects and snails.

Advantages of Azolla

1. It contains higher nitrogen content.
2. Release of nitrogen is easy.
3. There is no leaching loss.
4. Releases nutrient slowly.
5. Weed interference is less where grown as dual crop.

V. Blue green algae or Cyanobacteria

BGA is phototrophic in nature. These are photosynthetic, prokaryotic microorganisms and capable of fixing molecular nitrogen. BGA are found to grown in a paddy field and marshland. It is used as biofertilizer for irrigated rice crop. The rice field ecosystem provides an environment favorable for the growth of BGA with respect to their requirements *viz.*, light, water and temperature.

Nitrogen fixation by BGA have been estimated to be 25-30 kg N/ha. Yield increases by 1015 %. BGA has some species viz., *Nostoc*, *Anabaena*, *Olocia*, *Calothrix*, *Tolipothix etc*., They fix nitrogen in the presence of sunlight. They also contain growth promoting substances like auxin, Indole acetic acid and Gibberellic acid.

Recommendation: Application of 10 kg /ha flakes of BGA. It is applied 10 days after the transplanting of rice crop. BGA benefit the rice crop up to 25-30 kg N /ha.

Dose: Apply BGA @10-15 kg /ha after transplanting.

VI. Mycorrhize

Mycorrhizae means root fungi or VAM. These are obligate symbiotic. They are ubiqutous in nature and with very few exceptions found on a wide range of plant species. They form vesicular–arbuscular–mycorrhizal (VAM) symbiosis. VAM fungi infects and spreads inside the root by providing an aseptate mycelium. Vesicles which are sac like structure provided by these fungi are known to store phosphorus as phospholipids. Arbuscules formed inside the cortical cells are attributed to help in the exchange of nutrients. Common genera that form VAM associations are *Glomus*, *Gigaspora*, *Enterospora etc*.

Advantages

1. VAM fungi enhancing the nutrient uptake of P, Zn and S.
2. Enhances the water uptake under drought conditions.
3. Enhances resistance to root damage.
4. Increases the yield.

It is recommended for forest trees, forage grasses, maize, millets, sorghum and barely and also leguminous crops.

Mycorrhizae can substitute up to 560 kg phosphorous fertilizer per hectare for citrus. Only enriched soil in the presence of suitable plant has to multiplication are not yet known.

Phosphate solubilizing microorganisms

Phosphorus is present in soils in both organic and inorganic forms and most of which is insoluble and unavailable to plants. Several soil bacteria particularly those belonging to the genera *Pseudomonas* and *Bacillus* and fungi belonging to the genera Pencillium and Aspergillus possess the activity to bring insoluble phosphates in the soil to the soluble forms by secreting organic acids such as acetic acid, propeonic acid *etc*. These acids lowers the pH and bring about the dissolution of the bound phosphate forms.

Mode of applications: As that of VAM fungi

Constraints in usage of biofertilizers

1. Farmers acceptance has been for from satisfaction.
2. Difficult to demonstrate striking increase in yield due to biofertilizers unlike chemical fertilizers.
3. Strains fail to establish if the soil moisture in the surface zone.
4. Fungicide treatment is a problem in crops like groundnut.
5. Inoculants manufacturing is for below the requirement.
6. Availability of quality of inoculants is another problem.
7. Seed inoculation is no more suitable when seeds are sown on dry soil.
8. When seeds are coated with toxic chemicals and pesticides.
9. Soil containing higher number of ineffective native *Rhizobia.*
10. Seeds have fragile seed coat.
11. When seed coat is toxic to *Rhizobia.* In such situation soil inoculation with granulated inoculant is followed in USA, New Zealand, France *etc*., Granulated inoculation is prepared using grains of sand, gypsum, peat *etc*.

Chapter 18

Properties and Fate of Major and Micronutrients in Soils

Plant nutrients are naturally components of the soil orchestrated by weathering processes and that the abundance of these metals in our soil environment now-a-days is due to numerous anthropogenic activities. In the environment, heavy metals are partitioned among the various environmental compartments such as water, soil, suspended solids and biota. Processes such as dilution, dispersion, sedimentation, adsorption and desorption govern their distribution. They also exist in colloidal ionic particulate and dissolved phases, have a high affinity for humic acids, organic colloids and oxides coated with organic matter. The soluble forms are generally ions or unionised organo metallic chelates or complexes. The solubility of metals in soil and groundwater is predominantly controlled by pH, amount of metal, cation exchange capacity, organic carbon content, the oxidation state of the mineral components and the redox potential of the system. Soil pH, organic carbon, available amounts, oxides, effective cation exchange capacity, moisture content, oxidation-reduction state of the metals, ion exchange and sorption capacity for the metals are some of the soil factors influencing the fate and availability. At lower levels of these soil factors, their ions are rendered free in the solution thereby making them available to plant and underground water. In general, soil pH seems to have the greatest effect of any single factor in the solubility or retention in soils, greater retention and lower solubility of metal cations occurs at high soil pH. Both the concentration of trace metals and their speciation vary significantly, with the composition of soil solution and the amount of moisture content in the soil.

Chapter 19

Nutrient Toxicity and Management

The requirement of micronutrients is always in low amounts while their moderate decrease causes the deficiency symptoms and a moderate increase causes toxicity. In other words, there is a narrow range of concentration at which the elements are optimum. Any mineral ion concentration in tissues that reduces the dry weight of tissues by about 10 per cent is considered toxic. Such critical concentrations vary widely among different micronutrients. The toxicity symptoms are difficult to identify. Toxicity levels for any element also vary for different plants. Many a times, excess of an element may inhibit the uptake of another element.

Nutrient toxicities in crops are more frequent for manganese (Mn) and boron (B) than for other nutrients.

Manganese toxicity is found on acid soils. It is important to know that manganese competes with iron and magnesium for uptake and with magnesium for binding with enzymes. Manganese also inhibits calcium translocation in shoot apex. Therefore, excess of manganese may, induce deficiencies of iron, magnesium and calcium. Thus, what appears as symptoms of manganese toxicity may actually be the deficiency symptoms of iron, magnesium and calcium Fig 19.1 and 19.2).

Fig. 19.1: Boron toxicity symptoms

Fig. 19.2: Manganese toxicity symptoms

Boron toxicities occur in irrigated regions where the well or irrigation waters are exceptionally high in B. Most other nutrient toxicities occur when large amounts of nutrients in question have been added in waste, *e.g.*, sewage sludge. Crops grown near mines and smelters are prone to nutrient toxicities. Generally, the symptoms of toxicity in crops occur as burning, Chlorosis and yellowing of leaves. Toxicities can result in decreased yield and/or impaired crop quality.

Prevention of toxicity

1. With the exception of Mo, toxicity of other nutrients can be reduced by liming.
2. Following recommended rates of fertilizers and the safe and controlled use of waste materials, such as sewage sludge and coal fly ash, should reduce metal loading and nutrient toxicity in crops.
3. Use of crop species and genotypes less susceptible to toxicity are recommended where toxicity is suspected.
4. Provided sufficient drainage because availability of nutrients like Fe and Mn is increases up to toxicity level under water logged condition.
5. Ground water must be monitored regularly, if content of B and Cl is too high stop to applied water or applied with dilution.
6. Addition of sufficient amount of organic matter binds some of the toxic elements.
7. Ploughing in dry soil so increase the infiltration rate and leach the toxic element with rain water.

Chapter 20

Effect of Potential Toxic Elements in Soil Productivity

Toxic elements like Cadmium (Cd), Fluoride (F), heavy metals, radioactive metals are associated with mainly rock phosphates, other organic wastes like sewage sludge and pollutants. Their chemistry is similar to plant nutrients and replace them in their physiology leading to reduced uptake of nutrients, productivity and contaminated food chain.

Accumulation of toxic elements

Cadmium

The concentration of cadmium in phosphorus-containing fertilizers varies considerably and can be problematic. For example, mono-ammonium phosphate fertilizer may have a cadmium content of as low as 0.14 mg/kg or as high as 50.9 mg/kg. This is because the phosphate rock used in their manufacture can contain as much as 188 mg/kg cadmium. Continuous use of high-cadmium fertilizer can contaminate soil (as shown in New Zealand) and plants. Limits to the cadmium content of phosphate fertilizers has been considered by the European Commission. Producers of phosphorus-containing fertilizers now select phosphate rock based on the cadmium content.

Fluoride

Phosphate rocks contain high levels of fluoride. Consequently, the widespread use of phosphate fertilizers has increased soil fluoride concentrations. It has been found that food contamination from fertilizer is of little concern as plants accumulate little fluoride from the soil; of greater concern is the possibility of fluoride toxicity to livestock that ingest contaminated soils. Also of possible concern are the effects of fluoride on soil microorganisms.

Radioactive elements

The radioactive content of the fertilizers varies considerably and depends both on their concentrations in the parent mineral and on the fertilizer production process. Uranium-238 concentrations range can range from 7 to 100 pCi/g in phosphate rock and from 1 to 67 pCi/g in phosphate fertilizers. Where high annual rates of phosphorus fertilizer are used, this can result in uranium-238 concentrations in soils and drainage waters that are several times greater than are normally present. However, the impact of these increases on the risk to human health from radionuclide contamination of foods is very small (less than 0.05 mSv/y).

Other metals

Steel industry wastes, recycled into fertilizers for their high levels of zinc, wastes can include the following toxic metals: lead, arsenic, cadmium, chromium, and nickel. The most common toxic elements in this type of fertilizer are mercury, lead, and arsenic. These potentially harmful impurities can be removed; however, this significantly increases cost. Highly pure fertilizers are widely available and perhaps best known as the highly water-soluble fertilizers containing blue dyes used around households. These highly water-soluble fertilizers are used in the plant nursery business and are available in larger packages at significantly less cost than retail quantities. There are also some inexpensive retail granular garden fertilizers made with high purity ingredients.

Further Reading

Brady, N.C. and Weil, R.R. 2002. The Nature and Properties of Soils.13thEd. Pearson Edu.

Fageria, N.K., Baliger, V.C. and Jones, C.A. 1991. Growth and Mineral Nutrition of Field Crops. Marcel Dekker.

Havlin, J.L., Beaton, J.D., Tisdale, S.L. and Nelson, W.L. 2006. Soil Fertility and Fertilizers. 7th Ed. Prentice Hall.

Mengel, K. and Kirkby, E.A. Principles of Plant Nutrition. Kluver Academic Publishers.

Prasad, R. and Power, J.F. 1997. Soil Fertility Management for Sustainable Agriculture. CRC Press.

Russel's Soil Conditions and Plant Growth 13th Ed. –Allan Wild [ed.] English Language Book Society.

Yawalkar, K.S., Agarwal, J.P. and Bokde, S. 2000. Manures and Fertilizers. Agri-Horti Publ.

Other Publications on Horticulture

S.No.	Title	Author	ISBN	Year
1	A Colour Handbook on Practical Plant Pathology	Yadav, Vijay	9789385516177	2015
2	A Colour Handbook: Landscape Gardening	Singh, Alka & B.K. Dhakuk	9789383305889	2015
3	A Handbook of Minerals,Crystals,Rocks and Ores	Alexander, P.O.	9788190723787	2009
4	A Handbook of Soil-Plant-Water-Fertilizer and Manure Analysis	Durai, M.V.	9789381450185	2014
5	A Handbook on Irrigation and Drainage	Panigrahi, Balram	9789381450888	2013
6	A Text Book of Seed Science and Technology	Padmavathi, S.	9789381450437	2012
7	Abiotic Stress and Physiological Process in Plants	A. Bhattacharya	9789385516696	2017
8	Abiotic Stress and Plant Physiology *(in 2 parts)*	A. Bhattacharya and Luxmi Vijaya	9789385516900	2018
9	Abiotic Stress Tolerance in Crop Plants: Breeding and Biotechnology	Roy, Bidhan	9788189422943	2009
10	Acid Soils: Their Chemistry and Management	Sarkar, A.K.	9789381450383	2013
11	Advances and Challenges in Agricultural Extension and Rural Development	Rathakrishnan, T.	9789380235035	2009
12	Advances in Agri-Management	Rakhi Gupta *et al.*	9789385516894	2017
13	Advances in Fruits and Vegetables Processing and Preservation	V.K.Joshi		2018
14	Advances in Nutrient Dynamics in Soil-Plant Systems for Improving Nutrient use Efficiency	Elanchezhian, R.	9789385516962	2017
15	Advances in Preservation and Processing Technologies of Fruits and Vegetables	Rajarathnam, S.	9789380235523	2011
16	Advances in Protected Cultivation	Singh, Brahma, Balraj Singh	9789383305179	2014
17	Advances in Soil Borne Plant Diseases	Naik, Manjunath	9788189422813	2008
18	Agricultural Biotechnology: Indian Print	Persley, G.J.: et al	9788189422097	2006
19	Agricultural Extension: Worldwide Innovations	Saravanan, R.	9788189422967	2008
20	Agricultural Marketing: Perspectives and Potentials	Bhat, Anil & S.P. Singh	9789385516153	2016
21	Agricultural Plant Biochemistry	Nagaraj, G.	9789383305551	2015
22	Agricultural Statistics: A Guide for Competitive Examinations	Kushwaha, K.S.	9789381450314	2012
23	Agriculture Bioinformatics	Keshavachandran, R.	9789383305421	2015
24	Agri-Food Crops: Processing,Value Addition, Packaging and Storage	R.Sasi Kumar	9789381450406	2012
25	Agri-Horticultural Biodiverstiy of Temperate and Cold Arid Regions	Zeerak, Nazir Ahmad	9789381450086	2012
26	Agrobiodiversity and Sustainable Rural Development	Soam, S.K. & M.Balakrishnan	9789383305896	2015
27	Agro-enterprises For Rural Development and Livelihood Security	Sharma J.P. & S.K.Dubey	9789380235851	2011

For details, please visit us at www.nipabooks.com

28	Agroforestry for Increased Production and Livelihood Security	Sushil Kumar Gupta Pankaj Panwar & Rajesh Kaushal	9789385516764	2017
29	Agroforestry Systems for Resource Conservation and Livelihood Security in Lower Himalays	Panwar,P & Dadhwal	9789381450215	2012
30	Agroforestry: Principles and Practices	Patra, Alok Kumar	9789381450765	2013
31	Agroforestry: Systems and Practices	Puri, Sunil & Pankaj Panwar	9788189422622	2007
32	Agroforestry: Systems and Prospects	C.B.Pandey & O.P.Chaturvedi	9789381450970	2014
33	Agro-Informatics	Vanitha, G & M.Kalpana	9789380235714	2011
34	Agronomy: Principles and Practices	E. Somasundaram & M. M. Amanullah	9789385516740	2017
35	Agrotechnology Manual: Including Nursery Management and Practices	Etomes, Marcel N	9789383305018	2014
36	An Introduction to Intellectual Property Rights	Pathak, Manju	9789383305124	2014
37	An Introduction to Nanotechnology	Rathinasamy, A.	9789381450413	2012
38	Ancestral Knowledge in Agri-Allied Science	Saha, Ratan Kumar	9789383305216	2014
39	Applied Computational Biology and Statistics in Biotechnology and Bioinformatics (Set of 2 Vols.)	Roy, A.K.	9789380235929	2012
40	Applied Statistical Techniques	Imam, Ekwal	9789383305537	2015
41	Applied Statistics for Agricultural Sciences	Venkatesan, D.	9789383305285	2014
42	Approaches for Incorporating Drought and Salinity Resistance in Crop Plants	Chopra,V.L. & R.S.Paroda	9789383305742	2015
43	Aromatic Plants: Vol.01. Horticulture Science Series	Skaria, B.P. et.al.	9788189422455	2007
44	Auddaniki ke Adharbhoot Sidhant Tatha Nashijeev Prabandhan	Tiwari, A.K.	9789380235998	2012
45	Bananas and Plantains: Postharvest Management, Storage,Ripening and Processing	Narayana, C.K.	9789383305452	2015
46	Basic Concepts in Statistics	Kushwaha, K.S.	9788189422400	2009
47	Basics of Horticulture: 2nd Revised and Expanded ed. (As per Revised ICAR Syllabus)	Peter, K.V.	9789383305735	2015
48	Basics of Mutation Breeding	Thirugnanakumar, S.	9789383305193	2014
49	Basics of Research Methodology	Imam, Ekwal	9789383305438	2015
50	Biochemical Aspects of Plant Physiology: Technology and Methodology	Bhattacharya, Amitav	9789383305902	2016
51	Biochemistry,Molecular Biology and Biotechnology: Instant Notes	Gajera, H.P.	9789383305520	2015
52	Bioinformatics in Agriculture: Tools and Applications	Balakrishnan, M. & Dam Roy ed.	9789381450925	2014
53	Biometrical Methods in Horticultural Science	Nirmal Sharma, V.K. Wali, Parshant Bakshi	9789385516504	2016
54	Biostatistics: Basic Concepts and Methodology	Kushwaha, K.S.	9789383305469	2014
55	Biotechnology in Horticulture: Methods and Applications	Peter, K.V.	9789381450918	2013
56	Biotechnology in India: Initiatives and Accomplishments	Niladri, Bag	9789385516252	2016
57	Biotechnology: Practical Manual Series Vol 04	Thara, K.M.	9788190851237	2009
58	Bougainvillea	S.K. Datta, R. Jayanthi & T. Janakiram	9789385516702	2017

59	Breeding and Biotechnology of Flowers: Set of 2 Vols. (Set Price)	Singh, A.K.	9789383305612	2014
60	Breeding and Biotechnology of Flowers: Vol 01: Commercial Flowers	Singh, A.K.	9789383305353	2014
61	Breeding and Biotechnology of Flowers: Vol 02: Garden Flowers	Singh, A.K.	9789383305407	2014
62	Breeding and Protection of Vegetables	Rana, M.K.	9789380235493	2011
63	Breeding of Horticultural Crops: Principles and Practices: 2nd Revised & Expanded ed.	Kumar, N.	9789383305773	2015
64	Breeding,Biotechnology and Seed Production of Field Crops	Roy, Bidhan	9789381450680	2013
65	Cashew Production and Processing Technology: Recent Advances	Gajbhiye, R.C. & B.V. Padhiar	9789383305827	2015
66	Climate Change and Agricultural Food Production	Kibria, Golam et.al.	9789381450512	2013
67	Climate Change and Agroforestry: Adaptation, Mitigation and Livelihood Security	C.B.Pandey, M.K.Gaur & R.K.Goyal		2018
68	Climate Change and Food Security	Datta, M.& N.P.Singh	9788189422387	2008
69	Climate Change and Natural Resources Management	Lenka, & Lenka	9789381450673	2013
70	Climate Change and Plantations in the Humid Tropics	GSHLV Prasada Rao and C.S. Gopakumar	9789385516368	2016
71	Climate Change and Sustainable Agriculture	P Suresh Kumar	9789385516726	2017
72	Climate Mitigation and Carbon Finance: Global Initiatives & Challenges	Sahoo, A.K.	9789381450024	2012
73	Climate Resilient Animal Agriculture	GSLHV Prasada Rao		2018
74	Climate Resilient Crops for the Future	Peter, K.V.	9789383305599	2015
75	Climatic Variability: Impacts on Agriculture and Allied Sectors	Datta, M.:Ed.	9789381450949	2014
76	Commercial Crops Technology: Vol.08. Horticulture Science Series	Kurian, A. & K.V.Peter	9788189422523	2007
77	Commercial Horticulture	Patel, N.L. & S.L. Chawla ed.	9789385516238	2016
78	Computers in Agriculture	M. K.Sharma & Anil Bhat	9789385516160	2017
79	Crop Diseases Management: Principles and Practices	Narayanasamy, P.	9789380235677	2011
80	Crop Diseases: Identification,Treatment and Management	Darwin, Henry	9789380235462	2011
81	Developments in Physiology,Biochemistry and Molecular Biology of Plants Vol 01	Bose, Bandana & A.Hemantaranjan: ed.	9788189422028	2005
82	Developments in Physiology,Biochemistry and Molecular Biology of Plants Vol 02	Bose, Bandana & A.Hemantaranjan: ed.	9788189422929	2008
83	Dimensions of Extension Education	Mohapatra, B.P.	9789381450987	2014
84	Diseases of Field Crops and Their Integrated Management	Sanjeev, Kumar	9789385516283	2016
85	Diseases of Horticultural Crops Identification and Management: With Colour Illustrations	Kumar, Sanjeev	9789383305643	2015
86	Diseases of Vegetable Crops and Their Integrated Management:A Colour Handbook	Mishra, R.K.	9789381450499	2013
87	Drip and Sprinkler Irrigation	Biswas, R. Kumar	9789383305766	2015
88	Droughts in Agricultural Production: Monitoring & Management	Rao, G.G.S.N.	9789385516009	2015
89	Drug Discovery and Development: Traditional Medicine and Ethnopharmacology	Patwardhan, Bhushan	9788189422295	2007

90	Drying Technologies for Foods: Fundamentals and Applications	Prabhat K. Nema Arun S.Mujumdar	9789383305841	2015
91	Drying Technologies for Foods: Fundamentals and Applications: Part-II	Prabhat K. Nema	9789385516399	2016
92	Eco Agriculture Revolution	M.H.Mehta		2018
93	Ecologically Based Integrated Pest Management (Set of 2 Vols.)	Abrol,D.P.& U.Shankar	9789380235950	2012
94	Economics,Marketing and Sales of Agricultural Products	Thakur, S. Nath	9789381450659	2013
95	Efficiency Indices for Agriculture Management Research	Devasenapathy, P. & B.Gangwar	9788189422882	2008
96	Elementary Agriculture	Nainwal & Nainwal	9789385516535	2017
97	Elements of Entomology	H.Lewin, Devasahayam	9789381450635	2014
98	Emerging Technologies of the 21st Century	Roy, A.K.	9789383305339	2015
99	Encyclopedia of Agricultural Meteorology	Dhaliwal, L.K. & S.S.Hundal	9789380235547	2011
100	Enhancing Nutrient Use Efficiency	Ramesh, K	9789385516733	2017
101	Entomology: Novel Approaches	Jain,P.C. & M.C.Bhargava	9788189422325	2007
102	Essence of Horticulture	M.S. Patil, A.R. Karale,	9789385516481	2016
103	Essential Oils and their Applications	Das, Kuntal	9789381450741	2013
104	Ethnomedicinal Plants Resource of Orissa Vol 01	Sahoo, A.K.	9789380235752	2011
105	Evaluation and Impact Assessment of Technologies and Developmental Activities in Agriculture,Fisheries and Allied Fields	Roy, A.K.	9789380235400	2011
106	Experimental Biotechnology: Practical Manual Series 06	Dutta, Sunita & Abhijit Dutta	9789380235721	2011
107	Experimental Phytochemical Techniques	Raaman, N.	9789380235943	2012
108	Extension Management in the Information Age Initiatives and Impacts	Philip, H. & T.Rathakrishnan	9789381450543	2013
109	Extension of Technologies: From Labs to Farms	Anandaraja, N.	9788189422837	2008
110	Family Farming and Rural Economic Development	M.L.Choudhary & Aditya	9789383305858	2015
111	Flower Crops: Cultivation and Management	Singh, A.K.	9788189422356	2006
112	Flowering Trees: Vol.12. Horticulture Science Series	Valsalakumari, P.K.	9788189422509	2008
113	Flowers for Trade: Vol.10. Horticulture Science Series	Sheela, V.L.	9788189422516	2008
114	Food and Nutritonal Security by Sustainable Agriculture:	Mishra, Kumar Bijesh	9789383305049	2014
115	Food Engineering and Technology	Sharma, H.K. & Ashutosh Upadhyay	9789383305483	2015
116	Food Process Engineering and Technology	Pare, Aakash & Mandhyan	9789380235431	2011
117	Food Processing Waste Management: Treatment and Utilization Technology	Joshi V.K. & Satish Sharma	9789380235592	2011
118	Food Product Development and Process Innovations (in 2 parts)	H N Mishra		2018
119	Food Science	Bawa, A.S. & O.P.Chauhan	9789381450147	2013
120	Food Science and Technology: Glossary of Preeminence	Dev Raj	9789380235806	2011
121	Forest Seed Science and Management	Shukla, Gopal	9789385516757	2017
122	Forestry Science: Fundamentals and Terms	Sharad Nema	9789385516382	2016

No.	Title	Author	ISBN	Year
123	Fruit Breeding	Dinesh, M.R.	9789383305513	2015
124	Fruit Crops: Vol.03. Horticulture Science Series	Radha, T. & Lila Mathew	9788189422462	2007
125	Functional Foods	Mishra, H.N. Rajesh Kapur	9789383305988	2016
126	Functional Foods and Nutraceuticals: Sources and their Developmental Techniques	Riar, C.S. & Saxena, D.C. Ed.	9789383305964	2015
127	Fundamentals of Garden Designing: A Colour Encyclopedia	Roy, Rup Kumar	9789381450307	2013
128	Fundamentals of Ornamental Horticulture and Landscape Gardening	Tiwari, A.K.	9789381450079	2012
129	Fundamentals of Vegetable Production	Rana, M.K.	9789380235707	2011
130	Genetic Diversity and Phenotypic Stability in Crop Plants	S. Thirugnanakumar	9789385516955	2018
131	Genetic Resources and Seed Enterprises: Management and Policies: In 2 Parts	Ram, Hari Har & R.Yadava	9788189422653	2007
132	Genetics and Breeding of Flower Crops *(in 2 parts)*	M.S.Patil	9789385516948	2018
133	Genomics and Genetic Engineering	Satya, Pratik	9788189422776	2007
134	Geographic Information System	Gurugnanam, B.	9788190851282	2009
135	Geoinformatics Applications in Agriculture	Singh, Anil Kumar & U.K.Chopra	9788189422233	2007
136	Geospatial Technologies for Natural Resources Mgt.	Soam,S.K. & P.D. Sreekant	9789381450802	2013
137	GIS: Fundamentals,Applications and Implementations	Elangovan, K.	9788189422165	2006
138	Glimpes of Practical's in Extension Education	K. Pradhan		2018
139	Good Management Practices for Horticultural Crops	M.K. Jatav, *et al.*	9789385516641	2016
140	Guava: Evaluation Studies of Most Promosing Cultivars	Dolkar, Disket	9789385516849	2016
141	Heterosis Breeding in Vegetable Crops	Rai, Nagendra	9788189422035	2006
142	Horticulture Science Series Vol 01-12: Set of 12 Vols	Peter, K.V.: Series Ed.	9788189422479	2008
143	ICTs for Agricultural Extension: Global Experiments, Innovations and Experiences	Saravanan, R. ed.	9789380235240	2010
144	ICTs for Transfer of Technologies: Tools and Techniques	Verma, S.R.	9788193014479	2015
145	Identification and Management of Horticultural Pests	Ranjith, A.M.	9789381450567	2013
146	Illustrated Dictionary of Entomology	Paras Nath	9788189422561	2007
147	Illustrated Dictionary of Floriculture and Landscaping	M. Kannan		2018
148	Illustrated Plant Pathology: Basic Concepts	Darwin, Henry	9789380235080	2009
149	Improving Productivity of Drylands by Sustainable Resource Utilisation and Management	Dayal, Devi	9789385516191	2016
150	Indigenous Medicinal Plants and their Practical Utility	Lakshman, H.C.	9789381450116	2012
151	Information Access in Digital Libraries	Stanley, Madan Kumar	9789383305384	2014
152	Information and Communication Technology for Agriculture and Rural Development	Saravanan, R.	9789380235882	2011
153	Information and Knowledge Management: Tools, Techniques and Practices	Roy, A.K.	9789381450628	2013
154	Innovations in Food Processing Technologies	Saiti		2018
155	Innovations in Horticultural Sciences	Peter, K.V.	9789385516344	2016
156	Innovative Horticulture	Arunkumar, K.	9788189422738	2008
157	Integrated Farming System Practices: Challenges and Opportunities	Nanda, Sankarsana	9789385516207	2016
158	Integrated Nutrient Management in Kinnow Mandarin	Bakshi, Manish	9789385516573	2016
159	Integrated Pest Management in the Tropics: In 2 Parts	Abrol, D.P.	9789385516115	2016
160	Intellectual Property Rights Demystified	Ramkumar, Mu.	9788189422875	2008

161	IPR: Drafting,Interpretation of Patent Specifications and Claims	Rathore, N.S.: Ed.	9789381450819	2013
162	Irrigation and Agricultural Drainage Engineering	Biswas, Ranajit Kumar	9789383305247	2015
163	Irrigation Systems Engineering	Panigrahi, Balram	9789380235387	2011
164	Laboratory Manual of Biochemistry: Methods and Techniques	Sengar, R.S.	9789383305025	2014
165	Laboratory Manual of Microbiology: Practical Manual Series: 05	Roy, A.K.	9789380235189	2010
166	Library Services in the Knowledge Web	Veeranjaneyulu & R.Mahapatra	9789381450192	2012
167	Management of Horticultural Crops: Vol.11 Horticulture Science Series: (Part-I & II Combined in 1 Binding)	Pradeepkumar, T.	9788189422493	2008
168	Managing Soil Health for Maximising Crop Productivity	Obeng, F.k, Avornyo,	9789385516788	2017
169	Mechanization of Cultivated Crops	Singh, Surendra	9789383305759	2015
170	Medicinal Plants: Vol.02. Horticulture Science Series	Kurian, A. & A.Sankar	9788189422424	2007
171	Merging Plant Breeding with Crop Biotechnology	Yasin, J.K.:Ed.	9789381450598	2012
172	Methods and Techniques in Plant Physiology	Bhattacharya, A. & Vijay Laxmi	9789383305506	2015
173	Microbial Biotechnology for Sustainable Agriculture, Horticulture and Forestry	Bagyaraj, D.J.	9789380235820	2011
174	Microbial Diversity and Functions	Bagyaraj, D.J. & Tilak, K. et.al.	9789381450109	2012
175	Microbial Diversity and Its Applications	Barbudde, S.B., R.Ramesh & N.P.Singh	9789381450666	2013
176	Mobile Phones for Agricultural Extension	Saravanan, R.	9789383305230	2014
177	Modern Biotechnology and Its Applications (Set of 2 Vols.)	Behera, K.K.	9789381450833	2013
178	Modern Methods in Plant Physiology	Srivastava, Girish Chand	9789380235011	2010
179	Modern Technologies for Sustainable Agriculture	Kumar, Sunil	9789381450611	2013
180	Modern Technology in Vegetable Production	Hazra, Pranab	9789380235325	2011
181	Molecular Biology and Biochemistry	Puttaraju, H.P.	9788189422257	2008
182	Molecular Markers and Plant Biotechnology	Tomar, R.S.	9789380235257	2010
183	Nanotechnology in Agriculture	Subramanian, K.S.	9789383305209	2015
184	Nanotechnology in Soil Science and Plant Nutrition	Adhikari, Tapan	9789381450789	2013
185	Nematology: Fundamentals and Applications	Jonathan, E.I.	9789380235141	2010
186	Novel Food Processing Technologies	Nanda, Vikas & Savita Sharma	9789385516047	2017
187	Numericals and Short Questions in Farm Machinery, Power and Energy in Agriculture	Yadav, Rajvir	9788190723718	2009
188	Nutraceutical Values of Horticultural Products	Dhurendra Singh	9789385516979	2018
189	Nutritional Disorders in Fruit Crops: Diagnosis and Management	Prakash, M.	9789381450956	2013
190	Objective Agriculture	Rupinder Singh	9789385516993	2018
191	Objective Entomology	Devinder Sharma	9789385516290	2018
192	Objective Horticulture	Verma, Anil	9789381450239	2012
193	Oilseeds: Properties,Products,Processing and Procedures	Nagaraj, G.	9788190723756	2009
194	Orchid: Cultivation and Management	Tiwari, A.K.	9789383305001	2014
195	Organic Farming	Singh, A.K.	9789385516139	2015
196	Organic Farming: Scope and Uses of Biofertilizers	Panwar, JDS & Amit Kumar Jain	9789385516184	2016
197	Organic Spices	Parthasarathy, V.A	9788189422844	2008
198	Ornamental Gardening and Landscaping	M. Kannan, *et al.*	9789385516689	2016
199	Ornamental Horticulture	Sindhu,S.S.ed.	9789383305865	2016

No.	Title	Author	ISBN	Year
200	Ornamental Plants	Sabina, George	9788190851268	2009
201	Pest Management and Residual Analysis in Horticultural Crops	Gulati, Rachna & Beena Kumari	9789381450710	2013
202	Pesticides: Methods of Their Residues Estimation	Kumari, Beena & T.S.Kathpal	9789380235394	2010
203	Pests of Vegetables: Bionomics and Management	Laskar, Nripendra ed.	9789385516016	2015
204	Phal Evam Sabji Parirakshan Digdarshika	Kumari, Arunima	9789380235561	2011
205	Physio-Biochemistry and Biotechnology of Vegetable Crops	Rana, M.K.	9789380235318	2011
206	Physiological Disorders of Fruit Crops	Savreet, Sandhu & Bikramjit Singh Gill	9789381450581	2013
207	Phytochemical Techniques	Raaman, N.	9788189422301	2006
208	Plant Biochemistry: Techniques and Procedures	Nagaraj, G.	9789383305940	2015
209	Plant Diseases: Identification and Management (With Illustrations)	Rai, J.P.	9789383305315	2014
210	Plant Nutrient Disorders: Diagnosis and Management	Sarkar, A.K. & P. Mahapatra ed.	9789385516023	2015
211	Plant Pathogens and Principles of Plant Pathology	Kumar, Sanjeev	9789385516078	2016
212	Plant Secondary Metabolities	Shukla, Y.M. & R.Bhatnagar	9788190851220	2009
213	Plant Taxonomy and Biosystematics: Classical and Modern Methods	Rana, T.S.	9789383305414	2014
214	Plantation Crops	Bhani Ram	9789385516511	2016
215	Plants For Wellness and Vigour	V.L.Chopra	9789386546036	2018
217	Postharvest Management and Processing of Fruits and Vegetables: Instant Notes	Sharma, Satish	9789380235202	2010
218	Postharvest Techniques and Management for Dry Flowers	V.Ponnuswami & Aruna,P.	9789380235868	2011
219	Postharvest Technologies for Commercial Floriculture	Verma, Anil	9789381450048	2012
220	Postharvest Technology and Engineering: An Illustrated Guide	Dev Raj	9789381450451	2012
221	Postharvest Technology and Processing of Horticultural Crops	Pandit, P.S.	9789383305278	2014
222	Postharvest Technology of Horticultural Crops: Practical Manual Series Vol 02	Sharma, Satish & M.C.Nautiyal	9788190851206	2009
223	Postharvest Technology of Horticultural Crops: Vol.07. Horticulture Science Series	Sudheer, K.P. & V.Indira	9788189422431	2007
224	Practical Manual of Entomology (Insects and Non-Insects Pests)	Devasahayam,H.Lewin	9789380235905	2011
225	Practical Manual of Horticulture Crops: Set of 2 Vols.	Verma, Anil Kumar	9789383305711	2015
226	Practical Manual of Horticulture Crops: Vol 01: Production Technologies	Verma, Anil Kumar	9789383305704	2015
227	Practical Manual of Horticulture Crops: Vol 02: Processing and Postharvest Technologies	Vaidya, Devina	9789383305698	2015
228	Practical Plant Biotechnology and Genetics	Rani, Archana	9789383305995	2015
229	Precision Farming in Horticulture	Singh, Jitendar, S.K.Jain,L.K.Dashora	9789381450475	2013
230	Production Technology of Spices,Aromatic,Medicinal and Plantation Crops	Barche, Swati	9789385516061	2016
231	Propagation of Horticultural Crops: Vol 06 Horticulture Science Series	Rajan, S. & B.L.Markose	9788189422486	2007
232	Propagation of Horticultural Plants: Arid and Semi-Arid Regions	Singh, R.S. & R.Bhargava	9789383305254	2014

233	Protected Cultivation of Horticultural Crops	Singh, D.K & K.V.Peter	9789383305155	2014
234	Quality Control for Value Addition in Food Processing	Dev Raj, *et al.*	9789380235578	2011
235	Quantitative Genetics and Crop Breeding	Thirugnanakumar, S.	9789380235981	2012
236	Question Bank in Extension Education	Mohammad, Asif & K. Ponnusamy	9788193014493	2015
237	Question Bank in Forest Science	Gopal Shukla		2017
238	Question Bank in Fruit Science	Jarande, S. D	9789385516795	2017
239	Question Bank in Postharvest Technology	Guleria, SPS & Anil Kumar Verma	9789380235042	2010
240	Recent Advances in Biopesticides	Johri, Jayendra: eds.	9789380235219	2010
241	Recent Trends in Horticultural Biotechnology: In 2Vols	Keshavachandran, R.	9788189422592	2007
242	Rejuvenation in Fruit Trees: Tropical and Subtropical	Kumar, Rajesh	9789385516436	2016
243	Remember Your Humanity: Pathway to Sustainable Food Security	Swaminathan, M.S.	9789381450178	2012
244	Rural Livelihood and Food Security	Wani, M.H.	9789380235936	2012
245	Seed Processing: A Practical Approach	Rakesh C. Mathad	9789385516085	2018
246	Seed Production of Field Crops	Mondal, S.S.	9788190723763	2009
247	Seed Science and Technology	Vanangamudi, K.	9789383305117	2014
248	Seed Testing Techniques for Seed Spices	Sangeeta Yadav & Arun Kumar barholia	9789385516917	2018
249	Soil Conservation: Fully Revised and Updated: 3rd ed.	Hudson, Norman	9789383305971	2015
250	Soil Microbiology and Biochemistry	Hassan, G.Dar	9789380235134	2010
251	Soil Sampling and Methods of Analysis	Pal, Sushant	9789381450574	2013
252	Soil Science: An Elementary Textbook	Puri, A.N.	9789383305063	2015
253	Soil Testing and Analysis: Plant, Water and Pesticide Residues	Brajendra, Patiram	9788189422707	2007
254	Spices,Plantation Crops,Medicinal and Aromatic Plants: A Handbook	Tyagi, S.K.	9788193014486	2015
255	Spices: Vol.05. Horticulture Science Series	Nybe, E.V. and Mini Raj	9788189422448	2007
256	Statistical Designs and Analysis for Agricultural Field Experiments	Katyal, Vijay & D.M.Hegde	9789381450840	2013
257	Statistical Methods for Agricultural Field Experiments	Katyal, Vijay & B.Gangwar	9789380235424	2011
258	Sustainable Agriculture: A Vision for Future	Desai, B.K. and B.T.Pujari	9788189422639	2007
259	Sustainable Environmental Science	Sahu, D.D.	9789381450208	2012
260	Systematics of Fruit Crops	Sharma, Girish	9789380235066	2009
261	Tea: Technological Initiatives	Niladri Bag, Arundhati Bag and L.M.S. Palni	9789385516337	2016
262	Technologies for Sustainable Green Environment	Davamani, V.	9789381450420	2012
263	Temperate Horticulture: Current Scenario	Kishore, D.K.: et. al.	9788189422363	2006
264	Textbook of Floriculture and Landscaping	Singh, A.K.	9789386546005	2017
265	The Basics of Human Civilization: Food, Agriculture and Humanity Vol.01 Present Scenario	Prem Nath	9789381450734	2013
266	The Basics of Human Civilization: Food, Agriculture and Humanity Vol.02 Food	Prem Nath	9789383305377	2014
267	The Chemistry of Soil Processes	Greenland, D.J.	9789383305926	2015
268	The Pomegranate	Hiwale, S.S.	9789380235158	2009
269	The Qur-anic Plants and Animals	Zeerak, Nazir Ahmad	9789383305780	2015

270	The Science of Horticulture Vol 01	Peter, K.V.: eds.	9789380235479	2011
271	The Science of Horticulture Vol 02	Peter, K.V.: eds.	9789380235486	2011
272	The Theory of Sample Surveys and Statistical Decisions	Kushwaha, K.S. & Rajesh Kumar	9788189422899	2009
273	The Weeds of Kumaun Himalayan Region (Uttarakhand)	T.S. Rana & Bhaskar Datt	9789385516474	2016
274	Traditional Agricultural Practices: Applications and Technical Implementations	Rathakrishnan, T.	9789380235028	2009
275	Tuber and Root Crops: Vol.09. Horticulture Science Series	Palaniswami, M.S. & K.V.Peter	9788189422530	2008
276	Turning Plants Into Medicines: Novel Approaches	T, Parimelazhagan	9789381450468	2013
277	Underutilized and Underexploited Horticultural Crops: Vol 01	Peter, K.V.: ed.	9788189422608	2007
278	Underutilized and Underexploited Horticultural Crops: Vol 02	Peter, K.V.: ed.	9788189422691	2007
279	Underutilized and Underexploited Horticultural Crops: Vol 03	Peter, K.V.: ed.	9788189422851	2008
280	Underutilized and Underexploited Horticultural Crops: Vol 04	Peter, K.V.: ed.	9788189422905	2008
281	Underutilized and Underexploited Horticultural Crops: Vol 05	Peter, K.V.: ed.	9789380235288	2010
282	Value Addition in Flowers and Orchids	De, L.C.	9789381450123	2011
283	Vegetable Crops: Genetics Resources and Improvements	Singh,D.K. & H.Choudhary	9789380235509	2012
284	Vegetable Crops: Vol 04. Horticulture Science Series	Gopalakrishnan, T.R.	9788189422417	2007
285	Vegetable Seed Processing	Mathad, Rakesh C. & Basave Gowda	9789385516030	2015
286	Water Management for Enhancing Water Productivity	N.K.Gontia & H D Rank	9789385505728	2018
287	Weed Science	Das, P.C.	9789383305261	2015

FSC
www.fsc.org
MIX
Papier | Fördert gute Waldnutzung
FSC® C083411

Zeitfracht Medien GmbH
Ferdinand-Jühlke-Straße 7
99095 Erfurt, Deutschland
produktsicherheit@kolibri360.de